Security in Next Generation Mobile Networks: SAE/LTE and WiMAX

RIVER PUBLISHERS SERIES IN STANDARDISATION

Volume 4

Consulting Series Editor

Dr. Anand R. Prasad
NEC Corporation, Japan

Standardisation is a series addressing the pre-development related standards issues and standardized technologies already deployed. The focus of this series is also to examine the application domains of standardised technologies. This series will present works of foras and standardization bodies like IETF, 3GPP, IEEE, GISFI, ARIB, TTA, CCSA, WiMAX, Bluetooth, ZigBee etc.

Other than standards, this book series also presents technologies and concepts that have prevailed as de-facto.

Scope of this series also addresses prevailing applications which lead to regulatory and policy issues. This may also lead towards harmonization and standardization of activities across industries.

For a list of other books in this series, see final page.

Security in Next Generation Mobile Networks: SAE/LTE and WiMAX

Anand R. Prasad
NEC Corporation

Seung-Woo Seo
Seoul National University

River Publishers
Aalborg

Published, sold and distributed by:
River Publishers
PO box 1657
Algade 42
9000 Aalborg
Denmark
Tel.: +4536953197

www.riverpublishers.com

Cover designed by Alf Zugenmaier, Junko Prasad and Anand R. Prasad

ISBN: 978-87-92329-63-9

Dedication

To Akash, Ruchika and Junko

— Anand R. Prasad

To Sun-Won, Hyun-Wook, Hyun-June, and my parents

— Seung-Woo Seo

About the Authors

Anand R. Prasad, Ph.D. & Ir. (MScEngg), Delft University of Technology, The Netherlands, Certified Information Systems Security Professional (CISSP), Senior Member IEEE and Member ACM, is a NEC Certified Professional (NCP) and works as a Senior Expert at NEC Corporation, Japan, where he leads mobile communications related security activity. Anand is an active member of 3GPP SA3 (security standards). He is Member of the Governing Council of Global ICT Standardisation Forum for India (GISFI) where he also chairs the Green ICT group and founded the Security SIG. Before joining NEC Anand led the network security team in DoCoMo Euro-Labs, Munich, Germany, as a manager. He started his career as a researcher developing embedded solutions like MAC and ARQ for WLANs and later project leader of software modem team at Uniden Corporation, Japan. Subsequently he worked as systems architect (as distinguished member of technical staff) for IEEE 802.11 based WLANs (WaveLAN and ORiNOCO) in Lucent Technologies, The Netherlands, and as technical director at Genista Corporation, Japan, with the focus on perceptual QoS. Anand has also provided business and technical consultancy to start-ups, developed cost effective offshoring models and does business development in new markets. Anand has applied for over 30 patents, has published 6 books and authored over 50 peer reviewed papers in international journals and conferences. He is also active in several conferences as program committee member.

Seung-Woo Seo is the professor of Electrical Engineering in Seoul National University, Seoul and Director of Intelligent Vehicle IT (IVIT) Research Centre funded by Korean government and automotive industries. He received his Ph.D. from the Pennsylvania State University, University Park, USA, and B.S. and M.S. degrees from Seoul National University, Seoul, Korea, all in Electrical Engineering. He was with the Faculty of the Department of Computer Science and Engineering, Pennsylvania State University, and served as a Member of the Research Staff in the Department of Electrical Engineering in Princeton University, Princeton, NJ. In 1996, he joined the Faculty of Seoul National University, and since then he has served as Chair or a Committee Member in various international conferences and workshops including INFOCOM, GLOBECOM, PIMRC, VTC, MobiSec, Vitae, etc. His research areas include computer & network security, vehicular electronics and communication networks, and system optimization. Especially in the security area, he has been leading researches actively on IEEE 802.16 security, group key management for multicast and broadcast service, fast intrusion detection, and security economics. He has published numerous papers in the refereed journals and conferences, and has also published a book in Korean on security economics. He has served for five years as a Director of the Information Security Center in Seoul National University.

Preface

Mobile devices have witnessed an increasing demand in recent years. Such demand has been driven by the tremendous expansions in cellular networks, wide variety of mobile devices that meet all market segments and user needs, and the advancements in mobile services that allow both consumers and mobile worker alike to be more efficient and productive. As this trend continues to evolve over the next few years, and as more critical assets are added to the mobile device (e.g., mobile wallet, enterprise confidential data), security will continue to be one of the most important aspects of the network and device architectures. This book provides a fresh look at those security aspects, with main focus on the latest security developments of Third-Generation Partnership Project (3GPP), Evolved Packet Systems (EPS), and WiMAX. EPS is also known as SAE/LTE (System Architecture Evolution/Long-Term Evolution).

The first-generation of cellular systems (1G) lacked comprehensive security measures and protection. As those systems evolved, some limited security services began to be featured with them. With the evolution of the second-generation (2G) GSM systems, additional security measures were comprehended such as user-authentication and over-the-air data confidentiality. However, 2G systems still lacked mutual authentication and did not provide any protection in the core network of the cellular system.

The third-generation (3G) systems not only addressed these security aspects, but also addressed five security feature groups that are the network access security (how to provide users with secure access to 3GPP services and protect against attacks on the radio access link), network domain security (how to protect against attacks on the wire-line network and to allow nodes in the provider domain to exchange signaling data securely), user domain security (how to provide secure access to mobile stations), application domain security (how to allow secure exchange of messages between applications in the user and provider domains), and visibility and configurability of security (how to allow the user to observe whether a security feature is currently in operation and if any services depend on this security feature). 3GPP has been developing new standards with IP in the core network and with data rates up to 100 Mbps. This standard is also known as EPS, which specifies both the core aspects such the network Enhanced Packet Core (EPC) or System Architecture Evolution (SAE) and the access network aspects such as the Evolved-UTRAN (E-UTRAN) or Long Term Evolution (LTE). Early deployment of EPS is already available.

The goal of this book is to cover the EPS and WiMAX security aspects. The intended audience for this book is mobile network and device architects, designers, researchers, and students. The goal of the authors, who have a combined experience of more than 25 years in mobile security standardization, architecture, research, and education, is to provide the readers with a fresh and up-to-date look at the architecture and challenges of EPS and WiMAX security.

The book includes 6 chapters, where the first 3 chapters are intended to be introductory ones, and the remaining 3 chapters provide more in-depth discussions. The book starts with Chapter 1 where we give a background on Next Generation Mobile Networks (NGMN) activity and requirements. Following explanation of NGMN, Chapter 2 provides an overview of security, telecommunication systems, and their requirements. Chapter 3 provides some background on standardization. Chapter 4 discusses the EPS (or SAE/LTE) security architecture developed by 3GPP. In particular, this chapter covers the authentication and key agreement method for SAE/LTE together with newly defined key hierarchy. This chapter also addresses the challenging aspects of SAE/LTE interworking and mobility with UMTS together with the necessary key-exchange technologies. The focus of Chapter 5 is WiMAX (IEEE 802.16) security. Chapter 5 provides an in-depth discussion of the WiMAX security requirements, the authentication aspects of PKMv2, and the overall WiMAX network security aspects. In Chapter 6 we briefly cover security for (i) Home(evolved)NodeB (H(e)NB) is the Femto solution from 3GPP), (ii) Machine-to-Machine (M2M) security and (iii) Multimedia Broadcast and Multicast Service (MBMS) and Group Key Management.

Without any doubt, this book could not have been finalized had there not been valuable support from several people. It is our pleasure to be able to acknowledge those people whose help and advice shaped this book. Colleagues of the Department of Electrical Engineering in Seoul National University deserve special mention for their suggestions and comments on recent technology development. We are indebted to Young-Hoon Park, Dong-Hyun Je, and Seung-Tak Choi, graduate students in Seung-Woo Seo's laboratory, for helping in preparing the manuscript. Some parts of Chapter 6 were made possible with their precious contribution. We also thank Dr. Min-Ho Park in Samsung Electronics for providing documents on IEEE Standardization process. Our thanks also goes to Caroline Jactat and Xiaowei Zhang of NEC for reviewing and providing valuable comments on Chapter 4 regarding SAE/LTE security.

Anand Raghawa Prasad and Seung-Woo Seo
January 1, 2011

Contents

1

Introduction

Security threats and attacks are as old as human race. Security issues assume importance when they impact the everyday life of common people. The biggest impact on daily life happens when security threatens means to which life of common people is dependent. The mid-nineteenth century saw the beginning of computing, with a publication by Ada Byron that became a reality in 1935 as the electrical numerical integrator and calculator (ENIAC) [1]. Computing technology enhanced since then and in 1969 the Advanced Research Projects Agency Network (ARPANET) came into being connecting the computers and starting the era of the Internet [1]. The mid-nineteenth century also saw the birth of wireless communications in the form of Maxwell's electromagnetic (EM) wave postulates with first commercial mobile telephone system coming in operation in 1946 [2] while mobile connectivity to Internet started in mid-1990s [3]. Security attacks on computers, Internet, and mobile started in 1980s, with major issues appearing around the year 2000.

Today wireless technology and Internet have penetrated so deep in human society that many lives are dependent on it. To penetrate deep in human life, mobile communications has taken the path toward being open in terms of platform and technology used, and in bringing true Internet to the mobile terminal with its next generation mobile network (NGMN). The technologies making this happen are WiMAX [4] based on IEEE 802.16 [5] and third generation partnership project (3GPP) evolved packet system (EPS) [6]. Security if not provided properly in NGMN will be disastrous not only for the technology and standards but also for the life of several others. Thus we present security for NGMN in this book, with particular focus on EPS and WiMAX. In this chapter we first present NGMN and security requirements followed by an overview of rest of the book.

1.1 NGMN

We start this section with an overview of NGMN, followed by a brief description of security requirements as discussed by NGMN [9].

1.1.1 Overview

The NGMN project strives to bring recommendations and requirements for future mobile broadband communications. The activity provides operator view to the standardization bodies for the decade beyond 2010. The project looks into a target architecture based around the packet-switched core together with a new radio access technology. The expectation is that there is smooth migration from existing 2G and 3G networks to the future IP-based networks that should be both cost-competitive and provide broadband performance. NGMN thus brings mobile operator perspective that also looks into future business expectations.

From future business perspective, NGMN sees that customer expectation is on the rise that will require a double-digit factor increase in network capacity. Thus the user will be more demanding and expect to be satisfied while being agnostic to the technology being used and expecting ubiquitous service coverage. This also means that the business landscape will change with time and thus the resulting system should be able to cater for flexibility and scalability. While fulfilling customer demand, the cost aspect is of course also important for NGMN, which can be viewed in different ways: (1) simple economical viability of the new system, (2) the operating frequency to be used and maximizing of radio coverage/capacity, and (3) last but not the least, the ever-present legacy systems and migration path from these to new system or, in other words, the deployment aspect. Based on these, the NGMN project envisioned a roadmap, where the new system standardization is completed in 2008 and the trial solution available in 2009 with commercialization in 2010. Standardization of 3GPP SAE/LTE in this sense is reasonably on time and few operators, e.g., DOCOMO, Verizon, AT&T, and T-Mobile, are expected to commercialize their services in 2010–2011 timeframe, with others to follow

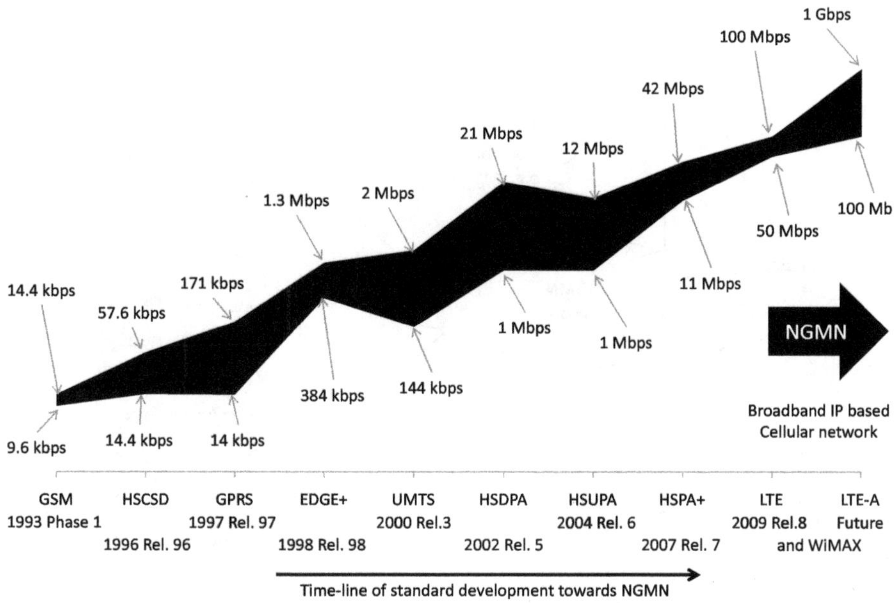

Fig. 1.1 NGMN and technology introduction (y-axis is log-scale for data rate).

by the year 2015. A high-level technology introduction roadmap as per NGMN is given in Figure 1.1. From these basic points of NGMN, one can see that both 3GPP EPS (or SAE/LTE) and WiMAX are in a position to fulfill the requirements.

NGMN also gives a high-level architecture that is in-line with 3GPP EPS and WiMAX (see Figure 1.2). The NGMN access network architecture is expected to consist of access nodes and access gateways. The network as such is expected to coexist with the circuit switched (CS) network that will phase-out with time. The expectation is that intelligence is moved to the edge and that there is provisioning of different business models. The solution is expected to maximize utilization of both the radio and PS core where the PS core will have a common PS anchor point providing transparent services over IP backbone that is capable of separating traffic based on quality of service (QoS) and security. The terminals are expected to fall back on legacy PS systems depending on coverage.

The expectation from NGMN project is summarized as following requirements, and more details are available in [8, 9]. These requirements

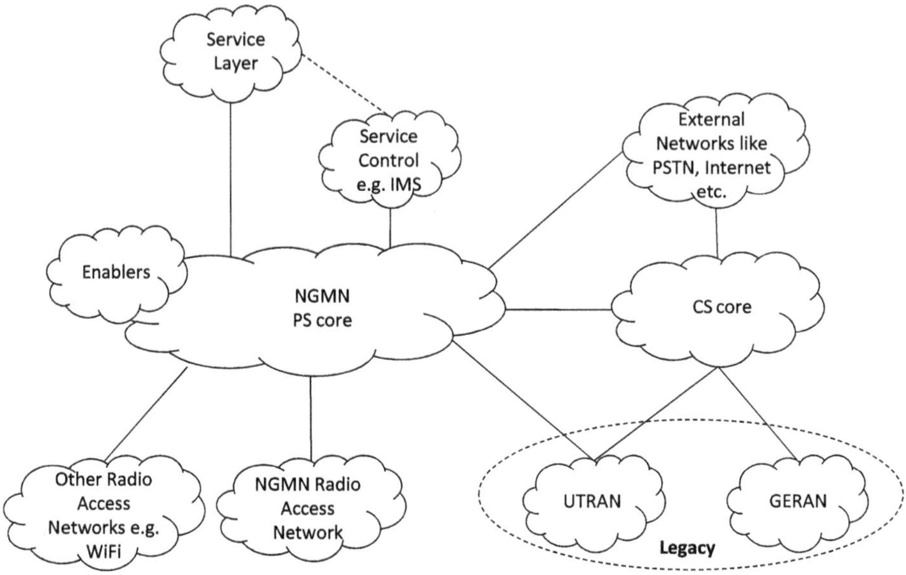

Fig. 1.2 High-level NGMN architecture.

can be categorized in three levels starting from (1) those that cannot be compromised, (2) those where some level of compromise could be allowed, and (3) where further comprise could be possible.

The following are the not-to-be compromised requirements:

- *Simplicity*: The system should have minimized complexity both in terms of architecture and protocols.
- *Total cost of ownership*: This has been discussed earlier, where the cost should consider migration, reuse of existing assets, upgrades, and operation.
- *Reliability*: Provisioning of sustained correct system operation.
- *Seamless mobility*: This is provisioning of service continuity during handover.
- *Spectral efficiency*: This means maximum reuse of available spectrum resources.

Requirements where some compromise is allowed:

- *Low latency*: This is the latency that is experienced by the user.

- *e2e throughput*: This is the actual goodput that is provided to the user.
- *QoS*: The expectation here is that predictable experience can be provided to users.
- *Security*: Here the need is to provide end-to-end security from device to service platforms. Security requirements are discussed in the following sub-section.

Requirements where further compromise is allowed:

- *Integrated network*: From NGMN perspective, integration is support of both NGMN and other access technologies.
- *Inter-working*: Here the requirement is put on coexistence with legacy networks.

Although not mentioned explicitly above, operations and management is an important aspect. One can observe that several of the discussions in this sub-section have impact on security.

1.1.2 Security Requirements

NGMN system requirement leads to a solution where IP for the first time will truly reach the mobile terminal that is supposed to be "always-on." Intelligence is moved to the edge, network architecture is flattened, and cost is decreased, all this for a new system that is expected to provide many new services. This means potential availability of true Internet in the mobile network, in contrast to IP as just transport. It is also expected that there will be self-organizing network functionality. All this leads to increased security issues, combined with open platform the overall security impact is far more than visible. It is strongly recommended that lessons learnt from Internet and Wireless LANs (WiFi) are taken in account from standardization to the product lifetime of NGMN system because today mobile networks are not only access means but also life-line of human society. In this subsection, we present NGMN requirements for security [9].

Security as per NGMN project is divided into a general part, access network security, service security, mobility security, and charging security. In general, NGMN expects that end-to-end security will be provided for protection from both attackers and unintentional damages. The expectation is also that security will be provided in a convenient manner to users and service providers.

From access network perspective, NGMN requirements are divided into two parts, one that is general and the other that applies to roaming cases. In general, for access network NGMN requires that (1) there is mutual authentication and authorization for subscriber and network; (2) provision usage of smart card for mutual authentication and authorization, and for service access; (3) tamper-proof identification of devices, useful for detection of stolen or non-compliant terminals; (4) provisioning of confidentiality of user and signaling traffic; (5) similarly to item (4), provisioning of integrity protection; (6) there should be secure means to measure and control the resource consumption; (7) prevention of network unlocking; and (8) limiting denial of service attacks (not including radio frequency (RF) jamming).

Looking at access network security from roaming perspective, the requirement demands (1) secure means of QoS provisioning; (2) allowing chargeable and authenticated/authorized access to subscribers from third party networks using their operator credentials; (3) secure routing of network signaling by third-party networks; (4) provisioning of secure traversal of IP traffic through third party network to/from operator core network; (5) fast security context transfer for handover between networks supporting same security functionality; (6) fast security transfer between access networks of two third-party networks with connectivity to the operator core network for the purpose of access security and (7) secure connectivity to operator core network over third party access networks that do not provide essential security.

The service security requires (1) provisioning of authentication and authorization for services provided by the operator; (2) confidentiality and integrity protection of signaling traffic related to services; (3) end-to-end security for user plane traffic with a key escrow mechanism for lawful interception; (4) provisioning usage of smart card for access to operator services with possibility of usage of other means like biometric; (5) portability across devices of service subscription credentials, third-party value-added service credentials and access keys; (6) service access independent of underlying bearer and access security; (7) secure identification of device as well as user to perform authorization; (8) different level of security provisioning for different session; and (9) single sign-on solution.

On the mobility side, the NGMN requires (1) secure intra and inter-system access mechanisms; (2) during intra access system handover the possibility of keeping the same security context without re-establishing a new one; (3) when creation of new security context is required the delay should not significantly affect real-time services; (4) secure handover instruction from network should be possible, and (5) provisioning of both user plane and signaling security during mobility.

In terms of charging, the requirement is that (1) integrity protection should be there to ensure that the operator charges the correct user for right service and (2) fraud protection throughout the system.

NGMN also requires reliability in terms of avoiding of single point of failure in the core network and fast and automated recovery from failures in a cost-effective manner. There is requirement on providing link and node failure recovery with end-to-end service recovery time of less than 1s.

1.2 Book Overview

In this chapter, we have given an overview of NGMN system and requirements. After this, we present two overview chapters so as to create a basis for the main chapters. To start with, we present security basics in Chapter 2 and an explanation of 3GPP and WiMAX standardization in Chapter 3. Following the basics, we present details of EPS security in Chapter 4 and WiMAX security in Chapter 5. As a conclusion of the book, we have collected certain topics related to current and future systems in Chapter 6. Topics covered in Chapter 6 are security for Femto, machine-to-machine (M2M) communication and mobile broadcast and multicast services (MBMS).

References

[1] B. Schell and C. Martin, *Webster's New World Hacker Dictionary*. Wiley Publishing, September 2006.

[2] H. Harada and R. Prasad, *Simulation and Software Radio for Mobile Communications*. Artech House, May 2002.

[3] GSM&EDGE, http://www.3gpp.org/article/gprs-edge, March 3, 2009.

[4] WiMAX Forum, http://www.wimaxforum.org/, March 3, 2009.

[5] IEEE 802.16, http://wirelessman.org/, March 3, 2009.

[6] 3G Americas, The Mobile Broadband Evolution: 3GPP Release 8 and Beyond HSPA+, SAE/LTE and LTE-Advanced, February 11, 2009.

[7] S. Aissi, N. Dabbous, and A. R. Prasad, *Security for Mobile Networks and Platforms*, Artech House, July 2006.
[8] NGMN Alliance, http://www.ngmn.org/, March 3, 2009.
[9] NGMN Alliance, Next Generation Mobile Networks. Beyond HSPA & EVDO, December 5, 2006.

2

Security Overview

The wide deployment of computers and Internet has caused a dramatic increase in information flow in the modern society. High-speed and reliable Internet access is possible through both wired and wireless media, thanks to the advancement of communication technologies. Moreover, with advances in multimedia software technologies linked with open application market (called application store), we are witnessing the opening of the new era of smart phones. However, together with penetration and enhancement of technology, security issues haunt these systems at unexpected instances. With security problems, it is difficult to predict how and when attacks occur and no individual or organization is free from the risk of attacks. Thus proper understanding of information security is necessary before developing security solutions for various systems. Before proceeding with chapters giving security overview of different technologies, in this chapter, we give an overview of information security and provide the essential concepts that will be used throughout this book.

2.1 Objectives of Information Security

The level of threat perceived by potential targets may differ depending on the value of their asset. This means the same kind of threat can be felt much amplified if the information asset is highly valuable. From the information security perspective, the value of assets determines what aspects of information can be exploited and thus must be protected against attacks. Some attackers may aim to acquire confidential information while others may want to replace certain information data or disable server operation. Similarly, defenders may require a certain type of assets to be confidential, while other type of asset to be authentic. Typically, attackers intend to achieve their goals by disabling a part or all of the following three major security objectives: confidentiality, integrity, and availability.

The objectives may vary from application to application and from institution to institution. For companies, the priority may be to prevent security incidents and minimize their impact in order to maintain business sustainability and reduce the loss. The three security objectives (explained in following subsections) are highly comprehensive, and many security-related international standards set similar security objectives accordingly.

2.1.1 Confidentiality

Confidentiality aims to maintain the secrecy of information by denying access to unauthorized persons. Confidentiality is typically achieved through encryption of data or systematic classification and management. Generally, the confidentiality of information is classified into several categories and its level is defined according to importance. For instance, the US military has five security levels: top secret, secret, confidential, restricted, and unclassified. As another example, there can be three security levels in a commercial domain: public, proprietary, and internal. Security level allows users to differentiate access rights to maintain appropriate security control. Nowadays, encryption algorithms have become quite secure and reliable; most of the vulnerabilities related to confidentiality result from other reasons, such as a defect in a communication protocol or key management.

2.1.2 Integrity

The purpose of integrity is to assure that the information has not been changed illegitimately and protect it from modification, deletion, and insertion. In general, the concept of integrity applies to not only the validation of the information itself (certification of content) but also the certification of the identity of a person who provides the information (certification of sender). That is because authentication of identity is often considered as a part of the validation process and is not separable from content certification. In a broader sense, the concept of integrity also includes the concept of a sender not being able to deny that he or she is the sender (nonrepudiation).

2.1.3 Availability

Availability refers to the property of continuous readiness of an object or resource without any delay when it is needed by authorized persons. For example, a server computer with availability must be accessed by authorized clients at any time. The denial of service (DoS) attacks against the availability of a network or server typically use a software method that exhausts the resource such as CPU or memory by exploiting the program errors. In many cases, availability is compromised by the secret operation of malware such as virus, Trojan horse, and worm. These programs often paralyze a major portion of a network by distributing traffic in a large scale.

2.2 Types of Security Attacks

In cyberspace, potential risk is manifested into cyber attacks through specific means. Most attacks use malicious programs, taking advantage of vulnerabilities of systems. Several international standards such as RFC2828 and X.800 list the techniques of cyber attacks. These attacks aim to compromise the three security objectives described previously, regardless of their specific means. In relation to the three security objectives, both computer and network have lots of inherent vulnerabilities against attacks. Computer security has vulnerabilities in terms of access right, account and user right, copyright, worm and virus, software reliability, and database. Although there is no clear distinction between computer security and network security nowadays, attacks exploiting various vulnerabilities in networks are rapidly increasing.

2.2.1 Types of Attacks on Networks

Security attacks on networks include a broad range based on diverse techniques. Some are direct types, i.e., altering or wiretapping through access to the confidential program/data stored in the remote location. They may rely on stealing user ID and password, masquerading, malicious replay, and analysis of communication pattern. On the other hand, some are indirect types that include interfering with service operation by taking advantage of system's weakness.

The following four types of attacks are common on networks: interception (or eavesdropping or wiretapping), modification, fabrication, and interruption. Interception compromises the confidentiality of security objectives, because it may lead to unauthorized access to data and analysis of traffic patterns. A malicious replay attack is committed by intercepting a document or message that has been already sent and resending it after certain duration of time. Modification is an attack against the integrity, including the act of modifying programs or data being transmitted. Fabrication also refers to an attack committed against integrity (or authentication) of security objectives including inserting specific communication session or data, or using fabricated user identity. Interruption is another attack performed against the availability by blocking the delivery of the whole or a part of information.

A DoS attack is a well-known type of the interruption attack. A DoS attack can be performed without degrading the crypto-related goals, e.g., confidentiality and integrity. As the most common type of network attack, its intention is to prevent an Internet site or service from functioning efficiently or at all, temporarily or indefinitely. In fact, since 2000, attacks have been attempted against such websites as Yahoo and eBay. At first, a DoS attack was used to target a single system or service. However, recently, a distributed denial-of-service attack (DDoS) has become more prevalent where multiple systems flood the bandwidth or resources of a targeted system, usually one or more web servers. This type of attacks uses multiple infected systems to perform its attack, which may result in compromising the entire network service.

The core part of network attacks is to pass the authentication procedure through impersonation. To impersonate a legitimate user, it is required to steal password through a password attack or wiretap to acquire user information. Spoofing is yet another form of deception used to impersonate specific identity, and IP spoofing is an attack where an attacker uses a false IP address to hide his or her identity. The man-in-the-middle (MitM) attack is a form of spoofing (similar to session hijacking) in which the attacker makes connections with the victims and relays false messages between them.

2.2.2 Security on Networks

With wide access to a network, a large percentage of activities on a computer are performed based on network connections. It is true that network connection provides people with unmatched convenience but it also exposes computers and its users to attacks. While it is not easy to provide complete security for computers, implementing security in a network connection is a far more difficult task. The reason why attacks through network are hard to defend is that attackers can hide their identity and specific/nonspecific multiple anonymous users can become either attackers or defenders. In networks, an anonymous attacker can make access through several hosts and network systems, and thus, disguise its origin or route. Meanwhile, various potential vulnerabilities may reside in a great number of different systems and application programs connected on the Internet. In addition, attacks on networks normally result in significant impact on many other systems, while attack tools and techniques are easily obtained in networks and spread out extremely fast through shared resources. Such attack tools are easy to use nowadays even by a layman.

2.3 Basic Security Functions

To counter cyber attacks on networks, many security technologies have been developed in both areas of computer and network security. The countermeasures cover a wide range of security areas, e.g., data encryption, authentication, access control, authorization, etc., whose importance becomes highly recognized as system/user applications in networks have become more complicated and user media have become diversified. It is well known that many of the security problems in networks go beyond the boundary of cryptology and cannot be solved simply by cryptographic knowledge.

So far, most of the security problems and countermeasures have been discussed for Internet environment. This is mostly because multiple vulnerabilities in the transmission control protocol (TCP)/Internet protocol (IP) stack has caused substantial security problems since it started operation in 1970. To make the usage of Internet secure, researchers have found that different security functions are necessary including authentication/authorization, data encryption, digital signature, nonrepudiation, and privacy guarantee. Furthermore, to use the public Internet as a private network by utilizing various security functions, they also proposed virtual private network (VPN). Thanks to the introduction of strong cryptography technologies such as public key infrastructure and digital authentication, current VPN offers superior security performance compared to an exclusive private network. The core technologies used in VPN are referred to as the tunneling technology based on IP security (IPsec) and secure socket layer (SSL), which are composed of many important basic security functions. Therefore, it is essential to understand the basic security functions that are used as building blocks of many security systems. We briefly give their overview in the following sections.

2.3.1 Authentication and Authorization

In a typical client–server network scenario, a client tries to access some system resources in a server. The client and server may be either fixed or mobile. In that situation, there are two important parts to controlling user access: authentication and authorization. Authentication refers to the method used to identify the subject requesting the permission to access information, while authorization defines the scope of actions allowed for a given system. Once the authentication and authorization are done, the client will have a right to access the specific resources. This right typically incorporates a set of parameters (called credentials), which are used throughout the rest of the communication. Following the credential negotiated during the authentication and authorization phase, the actual data communication may be protected by specific crypto algorithms. The protection is usually made by employing a broad set of security functions such as data encryption, integrity check, nonrepudiation, privacy protection, etc.

The granularity of authorization is dependent on the system and business, e.g., authorizing all mobile communications subscribers to access all services is very coarse and not good for business therefore authorization for different service or service set should be given. Also, authentication of just the client, or subject making service request, is not sufficient because the server providing service might be rogue. Therefore, server being authenticated by the client is necessary. Both client and server authentication to each other is known as mutual authentication.

2.3.1.1 Authentication

Authentication handles the problem of determining whether a client or server can be allowed access to a particular system resource. Classically, an entity can be authenticated based on the combination of the following three different elements:

1. Something you have: a physical token like a key
2. Something you know: something known only to you, e.g., a password
3. Something you are: some physical characteristic unique to you, e.g., biometrics.

To enhance accuracy, basic methods such as password or personal identification number (PIN) are combined with other types of technology such as hardware token, smart card, and public authentication certificate. Regarding the biometrics, although it is known that biometric authentication mechanisms work well where the person being authenticated is in presence, they work poorly over the Internet, because it is difficult to distinguish real authentications from replay attacks mounted by attackers who have captured the user's biometric information.

Thus, authentication over the network must be composed of two parts: one is how to provide securely the proof of authenticity for the information being delivered, i.e., integrity guarantees; the other is how to verify the proof of authenticity for the information, i.e., reliable identification. For example, for authentication by the fingerprint, it is not only necessary to verify that the fingerprint has been delivered without alteration and has been originated from a reliable source, but it is also necessary to make sure that the fingerprint received is the right one corresponding to the entity to be authenticated. This implies that authentication basically requires the ability to identify all targets such as users, processors, and information resources. Identification based on a simple name or a number is not enough in most cases, and typically relies on the information issued from the trusted third party. Authentication can be performed only under the premise that there is no error in the identification process.

Objects of authentication can be either clients (device or users) or messages. In client authentication, a client presents a set of credentials to the network and the network proves whether the credentials are really from the client. Although client authentication covers both device authentication and user authentication, it is not common for security designers to make explicit distinction between them [1]. It is because in traditional cellular phone communications, a user usually carries a device that contains a specific set of credentials created by network operators, and the credentials can be used to authenticate the user unless the device is stolen and used for illegal purposes. However, in modern communications where users can change the networks easily for different services, it becomes necessary to distinguish one from the other. Moreover, for company phones that may be shared by many users, user authentication should be done separately from device authentication.

(1) *User authentication*: Traditionally, user authentication has played a key role in remote access protocols such as the point-to-point protocol (PPP). Along with user authentication based on one of the authentication mechanisms (e.g., password authentication protocol [PAP] or challenge handshake authentication protocol [CHAP]), the protocol also configures some link-layer functionalities. During authentication, an authenticator requests a user to submit a password with other information such as credentials or challenges a user with a random number. Although PPP was widely used for its simplicity, some security and extensibility problems of the conventional authentication protocols have been raised, and then Internet Engineering Task Force (IETF) designed a new protocol called extensible authentication protocol (EAP) as an extension to PPP. Currently, EAP is considered the de facto standard authentication framework that supports diverse authentication mechanisms and allows the peers to choose a specific mechanism during the negotiation phase. Details of EAP are described in Section 2.5.

(2) *Device authentication*: When devices are authenticated instead of human users, public-key-based certificate or cryptographic keys have typically been used as credentials. In most wireless communications, these credentials are preloaded into devices either by manufactures or operators, and are not allowed to be accessed from outsiders. A well-known example of the standard certificate is X.509, which certifies the public key of the RSA encryption algorithm for the device. While device authentication usually precedes user authentication, it is made transparently to the user. In real operation, user authentication is typically performed once by

identifying a user when the credential is issued to the user, and the user has to carry it always in the form of a memorized password, a certificate, or a hardware token (e.g., a secret number card or a onetime-password generator).

(3) *Message authentication*: Contrary to the device authentication, the purpose of message authentication is relatively straightforward. It aims at ensuring that the information sent by the sender is not altered during transit as well as the sender is the trusted entity. (The second one is often called as the source authentication.) For message authentication, message authentication codes (MAC) based on cryptographic integrity checking algorithms, e.g., hash algorithms, are widely used. In typical communications, MACs are generated by the source device (or an application) and attached to the massage, and sent to the destination. When the destination receives the message with MAC, it verifies the message by recalculating the MAC and comparing it with the attached one. These codes are used for the methods such as retransmission after detecting the change in contents of a message. When a higher integrity level is demanded, MAC generated using the hash function (HMAC) needs to be attached to a message for authentication.

Authentication gives rise to many issues related to protocols especially when the entities are authenticated over the network. An authentication protocol, a kind of cryptographic protocol, defines a series of steps among two or more parties for accomplishing authentication. Following the milestone paper by Needham and Schroeder in 1978, a large number of authentication protocols have been proposed [2]. Although many of the protocols are unilateral (i.e., a server authenticates a client), mutual authentication is sometimes necessary in networks when the client wants to disclose its security information only to the real server. In mutual authentication, a client typically attempts to authenticate a server first, and if that succeeds, then the server authenticates the client.

Authentication has a long history, and a large number of specific mechanisms have been developed. However, in [3], the authors showed that the authentication mechanisms proposed so far are essentially similar, and they can be classified into only seven basic classes of authentication protocols. Their memo surveys the space of authentication mechanisms, describes the basic classes and provides examples of protocols which fit into each class. The seven classes are the authentication systems based on the following mechanisms:

1. Username/password
2. One time passwords
3. Challenge/response
4. Anonymous key exchange
5. Zero-knowledge password proofs
6. Server certificates plus user authentication
7. Mutual public-key authentication.

2.3.1.2 Authorization

Authorization refers to the technology that defines the scope of the operation allowed for a given system and provides continuous management. It determines whether an authenticated party has permission to access a particular resource or service. For example, a system administrator can authorize users only to read the files stored in a shared folder or grant them authority to edit and execute as well. Although authorization is often tightly bound to authentication, they are basically separate mechanisms. Authentication validates the identity of a party, while authorization defines whether the authenticated party can perform a certain action.

Authorization includes access control enforcement to the accessibility of a subject to the shared resources. The concept of authority in relation to the access control enforcement can be simply expressed in the form of a matrix. For example, when a row represents the list of all subjects and a column represents the list of all objects, its elements of the matrix represent the access rights of subjects. However, because search overhead dramatically increases with the size of such a matrix, the following methods are used instead of actual implementation. The first method is saving and managing an access control matrix for each column. Such collection of information is called access control list (ACL). On the other hand, access control enforcement can be managed for each row, which is referred to as "C-List" (compatibility list). Both ACL and C-List basically save the same information, but there are a few minor different aspects in their implementation. For example, in case of C-list, it is relatively easy to represent the relation between a subject and an object because the authorization is presented to the objects based on a subject. On the other hand, ACL indicates the authorization of subjects for an object and thus it is necessary to undergo an additional conversion procedure in order to identify the relation for a subject. However, the access control of a certain object is easily implemented on site for each system independently. Therefore, ACL is more widely used regardless of the fact that C-List has more advantages in application.

2.3.2 Message Encryption

In order to guarantee secrecy and privacy, the transmitted messages through communication channels are encrypted using the cryptographic technologies. It can be implemented using a cryptographic algorithm whose safety has been verified mathematically. Encryption algorithms can be classified into symmetric-key-based and public-key-based algorithms. Details regarding the encryption algorithms are described in the section on cryptographic primitives (see Section 2.4).

2.3.3 Digital Signature

Digital signature is a proof that a message with a specific signature generated by sender A was really originated from sender A. Basically, a digital signature can be considered as the opposite concept of the public-key infrastructure. A signature is an encrypted message generated by sender A using his personal key (also known as private key). No signature can be created by a person other than sender A since only sender A has the personal key. Sender A sends a message attached with the signature to receiver B who decodes the signature using sender A's public key and then performs verification by matching the decoded signature with the original message.

2.3.4 Nonrepudiation

A good application of digital signature on networks is nonrepudiation, which refers to a function that prevents a sender from denying its act of sending certain information after sending the information. It may also include a function that a receiver is prevented from repudiating its act of receiving the information.

It falls under the scope of integrity or accountability of the system, and the decision on this is made by a third party in the position with a fair view.

2.3.5 Privacy Guarantee

The concept of privacy in the digital world refers to its dictionary meaning, which is an individual's right to have his or her private life uninterrupted by others. For organizations, it means not to cause any disadvantages, inconveniences, or difficulties for individuals by collecting their private information (i.e., personally identifiable information) and distributing it to a third party. It is emphasized that even if personal information is collected legally, it needs to be managed safely. Privacy guarantee technology refers to the methods of establishing appropriate social, technical, and physical countermeasures to ensure the secrecy and anonymity of data.

2.4 Cryptographic Primitives

Cryptography is the most important core technology for information security in cyberspace. It enables numerous transactions in cyberspace to be performed in a secure manner. Even though cryptography technology is by no means perfect and requires numerous other assisting measures, it is definitely a necessity. The range of applications of cryptography technology is significantly wide, but it is mainly applied to confidentiality and integrity out of three security objectives. Depending on the level and user requirements of confidentiality and integrity in communication, cryptography technology can provide different services for various applications.

2.4.1 Symmetric-Key Cryptography Algorithm

A symmetric-key cryptography algorithm refers to one in which the same key is used for encryption and decryption. The key is a random bit stream of constant length. DES is the cryptography algorithm standardized in 1977, and is currently used in a large number of products in various fields of applications. All the internal operation principles of DES are known. However, different user groups or individual users can select their own key and thus disclosing the algorithm itself does not cause a security problem. For other types of symmetric-key cryptography algorithm, there are algorithms such as triple-DES, RC4, RC5, IDEA, Blowfish, AES, and so on.

2.4.2 Hash Function and MAC

The hash function and MAC are a type of cryptographic algorithms that verify the authentication and integrity of messages by receiver. As for authentication, they make sure whether a message came from a right sender. The hash function calculates a hash value in a single direction without using the key, while MAC uses a shared secret key during the calculation. A sender precalculates the MAC value for a message and transmits it as an attachment to an original message. A receiver then calculates the MAC value of the message using the same methods and compares the result with one received. In this way, if a third party intercepts the message and changes its contents or MAC, the receiver can identify whether the message or MAC has been altered during transmission. Such a method can also be applied to data stored in a local storage. For MAC, a symmetric-key cryptography algorithm such as DES can be used. As a standard hash mechanism, the HMAC framework was defined by IETF, and with a specific hash algorithm, HMAC is used in the form of HMAC-MD5 or HMAC-SHA1.

2.4.3 Public-Key Cryptography Algorithm

The public-key cryptography algorithm is an asymmetric-key cryptography method that resolves the key distribution problems in the symmetric-key cryptography. The concept of public-key cryptography was devised by Whitefield Diffie and Martin Hellman in 1976 based on the mathematical principle that the discrete logarithm problem into two prime numbers is extremely difficult to solve. Afterwards, many public-key cryptographic algorithms have been proposed, e.g., RSA, ElGamal Encryption, and Elliptic Curve Cryptography(ECC), based on different mathematical principles. While RSA is based on the principle that factorization of a large fixed number into two prime numbers is extremely difficult, ElGamal Encryption and ECC are based on the difficulty of the discrete logarithm problem.

2.4.3.1 Operation principle

Although all mathematical principles used in various algorithms are different, their working concept remains the same. In a public-key cryptography algorithm, each host generates two keys paired up as one at all times using the designated key generation algorithm. One of the keys is selected as a public key and the other one as a private key. The public key is known to all, while the private key is kept secret within the host. When a person needs to send a message to host A, it performs encryption using A's public key and sends the result to host A. It is not necessary for any communicating hosts to agree on the key with host A in advance nor need to know who host A is. When host A receives the message, it decrypts the message with the other key in his possession (private key).

To apply the public key algorithm, authentication of a third party organization is required to determine whether the public key of each individual is trustworthy. The system that defines all relevant procedures between the three parties including key management based on the public-key cryptography algorithm is referred to as public key infrastructure (PKI). The third party that issues and authenticates the public key of the communicating parties is referred to as certificate authority. Most important communications such as financial transactions and private information reporting are currently relying on PKI.

However, in reality, a large-sized message is not directly encrypted or decrypted using a public key because of the computation complexity of the algorithm. In a case where the computation overhead is large, they use PKI for delivering a session key for communication. Actual communication is based on the symmetric algorithm using the session key. Application systems such as PGP, PEM, and S/MIME use this kind of method.

2.4.3.2 PKI applications

As mentioned previously, PKI has other important applications, e.g., digital signature and nonrepudiation. Regarding nonrepudiation, there is no countermeasure with MAC if a sender himself denies sending the message because both sender and receiver can create the same MAC. However, nonrepudiation through digital signature solves this problem because only the sender can attach one's signature to a message.

In a case where it is actually implemented on networks, digital signature is not created by encrypting the message directly with a public key cryptography algorithm due to its performance (speed) issues and mathematical security problem.[1] Instead, a sender calculates the one-way hash value of a message

[1]For the detailed mathematical example in relevance, refer to Section 10.6 in [4].

and attaches its signature on the hash value using its private key. When a receiver receives the message attached with its signature, he calculates the one-way hash function of the original message and decrypts it with the public key of the sender to compare whether it matches the received signature.

2.4.4 Random Number Generator

Random number is an element used in almost all information security systems such as cryptographic algorithm, MAC, and digital signature. A random number generator is used to create a key or a nonce. Theoretically, it is not possible to create a completely random number; so actually, a pseudorandom number is used instead. The purpose of a pseudorandom number generator is to create as unpredictable or irreproducible number as possible. To create a random number, a random value known as "seed" is used as the initial value, which is mostly obtained in physical phenomenon, for example, the arrival time of a packet and the size of electromagnetic noise.

2.4.5 Key Management

The validity of cryptographic algorithms is dependent on proper management of keys. Cryptographic algorithms for confidentiality and integrity cannot work if keys have not been set up securely in advance. The key management issue is essential in both the symmetric-key and the public-key methods. Cryptographic key management includes such procedures as key generation, distribution, storing, updating, revoke, and their complete disposal.

2.4.5.1 Key exchange in symmetric-key cryptography

The biggest problem in symmetric key cryptography algorithms lies in managing the key. In order for a system to operate properly, a secret key needs to be shared before sending or receiving the messages between users. Instead of using the same key for every communications, two parties establish a new key, called a session key, whenever they want to start new secure communication. A session key should not be one used in the previous run of the protocol. A typical way to set up a session key is to rely on a trusted third party (or a server), which will be engaged in the key management protocol. As shown in [2], a vast number of key establishment protocols have been proposed so far. Considering the fact that all the messages should come from authenticated parties, the key establishment protocol is usually combined with an authentication protocol.

2.4.5.2 Key management in public-key cryptography

Key management in public-key cryptography is also done via a trusted third party. Once each communicating party generates a pair of public and private keys, it registers its own public key to the trusted third party. The public key of a communicating party is available in the third party along with the certificate verifying its authenticity. The certificate issued by the trusted third party not only contains the public key of a communicating counterpart but also various other information such as the user's name that possesses the key, when the signature was made on the certificate, when it expires and what algorithm a public key uses. A certificate is used in many applications on Internet, e.g., IPSec or SSL as well as in modern wireless communications. Note that the communicating parties need to make sure whether an organization that authenticates a certificate is also reliable because they may be exposed to "man-in-the-middle" attack.

2.4.6 Cryptographic Protocol

For secure communications, it is necessary to develop not only secure cryptographic algorithms but also a reliable cryptographic protocol. A cryptography protocol basically involves some cryptographic algorithms, aiming at achieving diverse cryptographic goals such as keeping secrecy, sharing secrets, jointly generating secrets, etc. Many well-known security systems, for example, S/MIME and OpenPGP for e-mail security, SSL, and TLS for web browser security, and Microsoft PPTP, L2TP, and IPSec for secure transmission of IP packets, are implemented on secure cryptographic protocols. Note that most of cryptographic protocols go usually with a key management protocol. For example, IKE is the protocol that takes charge of key exchange at IPSec.

2.4.7 Comments

The basic cryptographic technologies described so far can be applied to networks in two levels. One is a transmission link level and the other is an end-to-end level. Message encryption in a transmission link level implies that all the remaining parts (i.e., network-layer header, transport-layer header, application-layer header) except the link-layer header part within the frame structure of a message are encrypted. When this method is used, each intermediate node that is on the route to the destination needs to perform encryption and decryption for each frame. Key sharing can be limited to two adjacent nodes. Accordingly, link-level encryption has an advantage that it can prevent the analysis attack of the traffic pattern between source and destination. Moreover, it can reduce the total number of keys. However, the overhead in relation to the frame processing on internal nodes may be relatively large.

On the other hand, end-to-end encryption refers to the method of encrypting only the message payload that includes the application-layer header. End-to-end encryption can be considered as a sort of "logical link" since it is applied to the session between two application processors. Since end-to-end encryption is typically done in the user application program, it is relatively easy to implement the required functions in a flexible and selective manner. However, all nodes must exchange and manage the key separately from the other nodes. Therefore, when the number of nodes on a network increases, the number of required keys increases dramatically.

Comparing the two methods, there is no rule to use only one of them. Thus, based on the individual requirements, both methods can be used simultaneously in cases where the reliability of a certain link needs to be particularly strengthened along with the end-to-end security. The integrity of the transmitted information on a network may also be guaranteed in two levels. This integrity guarantee includes not only the integrity of the data itself but also that of the header information attached to a packet or a frame (e.g., network information such as source and destination address). Source authentication is expected to gain more attention because disguised attacks through networks have become more common than before.

2.5 Extensible Authentication Protocol

EAP is an IETF standard authentication protocol (RFC 3748) designed to run over data link layer protocols, e.g., PPP, IEEE 802.1X, or Ethernet. EAP is not a particular authentication mechanism but a framework for authentication. In other words, EAP does not actually perform authentication, but supports the peers so that they can negotiate as they like and provides an infrastructure for end entities to authenticate each other. EAP is becoming more and more popular because of its flexibility in deploying various and newer authentication mechanisms and algorithms without significant changes in devices. In this section, we review the highlights of EAP in some more detail.

2.5.1 General EAP Architecture

2.5.1.1 EAP elements

EAP has been developed based on the following elements (see Figure 2.1):

- *EAP peer*: An edge device that is attempting to connect to a network, also known as an access client. It may be separate from an end user who actually holds and uses the peer as a device.
- *EAP authenticator*: An edge device (e.g., access point in WLAN or network access server (NAS)) that requires EAP authentication prior to granting access to a network.
- *Authentication server*: An entity that actually authenticate the peers. It validates the EAP peer's credentials and authorizes access to the network. It works as an AAA (Authentication, Authorization and Accounting) server, and can be a remote authentication dial-in user service (RADIUS) server or a Diameter server.
- *EAP server*: An entity that terminates the EAP authentication method and other EAP-related functions.

2.5.1.2 EAP authentication models

EAP authentication can be implemented in two ways, i.e., two-party authentication and three-party authentication. In two-party authentication, the

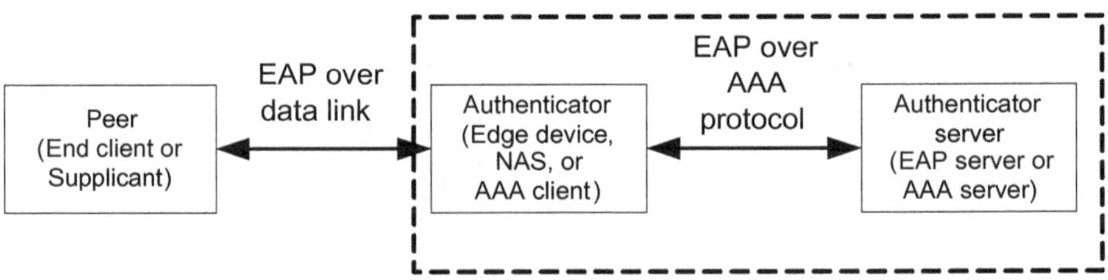

Fig. 2.1 Three-party authentication model.

authenticator authenticates the peer by itself, while in three-party authentication, it asks a server for authentication. In two-party authentication, an EAP server can be the EAP authenticator, while in three-party authentication, it can be the authentication server that negotiates the use of a specific EAP method with an EAP peer. As a three-party authentication protocol, a standard protocol, either RADIUS or Diameter, is typically used.

2.5.1.3 EAP authentication

When peers decide to run EAP during the connection authentication phase, they negotiate the use of a specific EAP authentication scheme known as an EAP method, allowing diverse plug-in modules for various authentication methods. This is depicted in Figure 2.2 that gives generic EAP message sequence, and shows different authentication methods in the square box can be applied.

To add support for a new EAP method, one can simply install an EAP method library file on both end entities, i.e., EAP peer and the authenticating server. After an EAP method is determined, EAP starts the exchange of messages between the end entities. The types of EAP messages are requests/responses for authentication information and success/failure. As message arguments, the length and details of the authentication are determined by an authentication method.

For three-party authentication, an EAP peer requests connection to a network through EAP authenticator. The authenticator receives an Identity message from the peer, which is transmitted to an authentication server via an AAA protocol, RADIUS, or Diameter. The EAP peer may be a supplicant, which is a software component that uses EAP to authenticate network access. The authentication server validates EAP peer's credentials and determines whether the peer is a valid one. Once authorized, the peer is connected to the network as requested. Note that since EAP messages are exchanged logically between the EAP components on an EAP peer and an authentication server, the EAP authenticator does not need to support any specific EAP methods.

2.5.2 EAP Authentication Methods

Using the type field of EAP request and response messages, it is possible to indicate a specific authentication method. Some of the methods popularly used are message digest 5 (MD5)-challenge, one-time password (OTP), generic

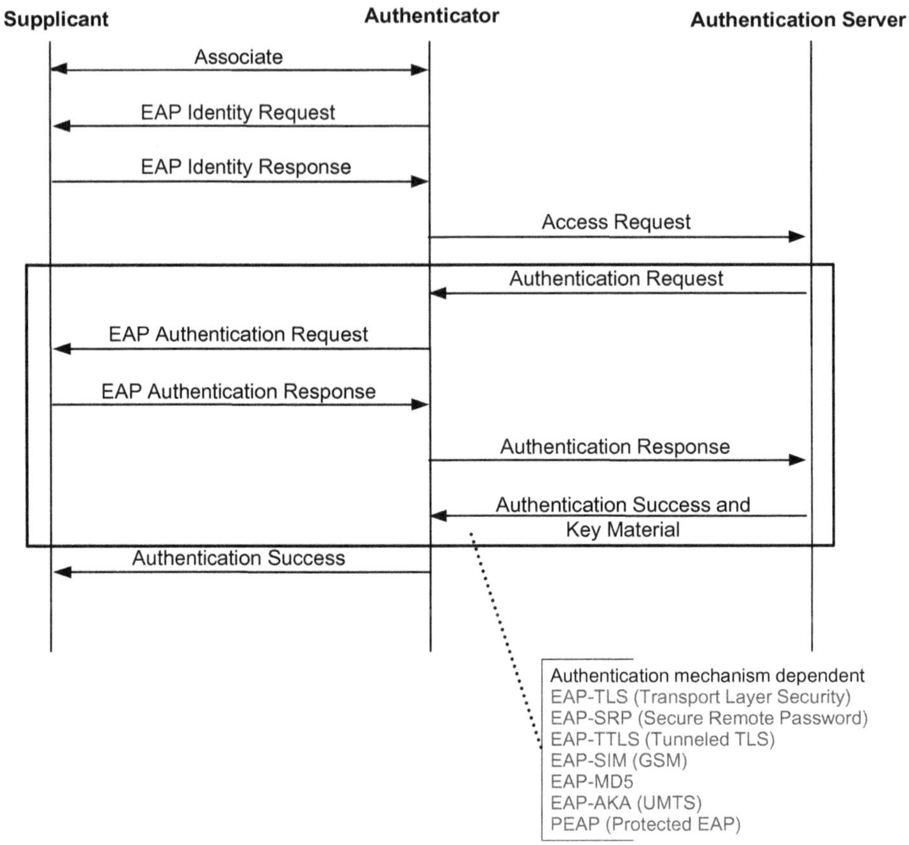

Fig. 2.2 High level message sequence for EAP.

token card, RSA public-key authentication, and transport layer security (TLS) for smart card and digital certificate-based authentication. Combined with EAP, they are called as EAP-TLS, EAP-TTLS (Tunneled TLS), Lightweight EAP (LEAP), and Protected EAP (PEAP), etc. In the following, we briefly discuss the features of two of them — EAP-TLS and EAP-TTLS — as representative examples.

2.5.2.1 EAP-TLS (TLS over EAP)

EAP-TLS takes the advantages of both TLS and EAP in the form of increased security and flexibility, respectively. EAP-TLS specification was developed to provide the functions of mutual authentication and key management mechanisms over PPP links, which were not supported in many of the other methods. EAP-TLS utilizes the features of TLS, which is composed of two layers: *handshake layer* that performs authentication, negotiating cipher suit and key exchanging, and *record layer* that provides end-to-end security protection at a transport layer. TLS messages are encapsulated within EAP packets, and due to the encapsulation, NAS does not need to take any actions for TLS handshake or record layer details.

TLS is regarded as a strong authentication method for the EAP framework because TLS can use digital certificates for identity verification. This feature is particularly helpful in wireless communications, where authentication is necessary between a roaming client and any visited network for secure communication.

Despite many features of EAP-TLS, it has some inherent drawbacks [1, 10]. First, since EAP-TLS is designed for the case where server and client both have certificates for mutual authentication, in case when only unilateral authentication is possible, it may not be applicable. Second, because EAP-TLS requires the user-level certificate, not the device-level certificate, it may incur large overheads to obtain the user certificate whenever authentication is carried out for a large number of subscribers. Note that certificate-based EAP typically requires a long series of message exchanges, which is not good for mobility. Third, in EAP-TLS, the entire EAP conversation is sent as clear text (unencrypted). Specifically, EAP-TLS does not provide any mechanism that can safely deliver the key negotiated between a peer and a server to the authenticator. With this problem, a malicious user can inject packets into the connection or capture the EAP messages from a successful authentication for offline analysis. This security issue is especially problematic for wireless connections.

2.5.2.2 EAP-TTLS

EAP-TTLS was developed to overcome the above drawbacks of EAP-TLS. EAP-TTLS inherits the characteristics and the generic model of EAP-TLS except for the following features. In EAP-TTLS, the server authentication to a client is done by using the server certificate as in EAP-TLS. However, client authentication is allowed to be done in a much simpler way by accommodating the less sophisticated legacy mechanisms such as PAP, CHAP, and MS-CHAP through RADIUS infrastructures. These mechanisms are operated on a secure TLS tunnel, which must be set up before the client authentication. The secure tunnel is necessary for wireless communication, and often used for security profile negotiation between client and server. Regarding the message format, TLS negotiation records such as specific mechanism-based authentication messages and key exchange messages are encapsulated within EAP-TTLS packets, which are again encapsulated within EAP packets.

Another important feature of EAP-TTLS compared to EAP-TLS is in roaming; EAP-TTLS provides a way for the authentication between the client and a new domain into which the client has moved. In this case, the new domain may not have trustful relationship with the client beforehand. In order to support this function, EAP-TTLS proposed to deploy another AAA server called TTLS server, with which the client establishes a secure TLS channel and exchanges all necessary messages. If the TTLS may not be able to authenticate the client by itself, it must refer to the home server of the client.

EAP-TTLS provides the support of faster negotiation if the client and TTLS server have already established a session previously. This feature is particularly useful for client mobility. Once a session has been established between a client and a TTLS server, it can be resumed later in a fast way if the server has not completely erased all the records on the session. Fast session resumption is possible by including the session ID in the first message of TLS handshake, i.e., a client hello message.

2.6 AAA Protocols

For the three-party model in networks with a large number of clients, we note that it is much helpful to have a separate back-end authentication server than to handle the client's requests at the network access point (or attachment point). Most of the back-end servers now have the full functions of authentication, authorization, and accounting (AAA) within a single server. Currently, as an AAA protocol, RADIUS is the most widely used, while in the future, Diameter is expect to substitute RADIUS. In this section, we summarize RADIUS and Diameter specifications, the full versions of which are shown in RFC 2865 [6] and RFC 3588 [7], respectively.

2.6.1 RADIUS

RADIUS was originally designed to work in a scenario where NAS relays the request of a dial-up user with its credentials to a backend server, and vice versa. So, the messages were defined to support PAP and CHAP. Later, it was extended to support EAP. One of the key features of RADIUS is that it is based on a client–server model, where a NAS works as a RADIUS client. The client is responsible for transferring end-user information to designated RADIUS servers, and then acting on the returned response. RADIUS servers perform such functions as receiving user connection requests, authenticating the end-user, and returning all configuration information that is necessary for the client to deliver service to the user. The RADIUS server is flexible in supporting a variety of methods to authenticate a user.

The general operation of RADIUS can briefly be described as follows. With the RADIUS configuration in a client, a user presents authentication information, e.g., username and password, to the client. Once the client has obtained such information, it may create an access-request message containing such attributes as the user's name, the user's password, the ID of the client, and the Port ID which the user is accessing, and send it to the RADIUS server via the network. When a password is present, it is hidden by using a method based on a cryptographic algorithm, MD5. In case a request from a client does not have a shared password, it must be dropped silently. If no response returns within some length of time, the request is resent a number of times.

The RADIUS server validates the sending client, and once the client is valid, the server looks up the user database to find the requester. Database lookup typically requires a series of matches, e.g., verification of the password and the specific port to which the user is allowed access. If all matches are successful and the RADIUS server determines to issue a challenge to which the user must respond, the RADIUS server transmits an Access-Challenge response. In this challenge/response authentication mechanism, the user is given a nonce (a random number) and requested to encrypt it and return the result. If the client receives an Access-Challenge and supports the challenge/response mechanism, the client then submits the second Access-Request (the original Access-Request with a new request ID and with the User-Password Attribute replaced by the encrypted response). The server can respond to this new Access-Request with one of the three messages: Access-Accept, Access-Reject, or another Access-Challenge. The configuration values for the user are placed within an Access-Accept response. These values include the type of service (e.g., whether the service is PPP), and other values necessary to deliver the desired service. For PPP, this may include values such as IP address, subnet mask, MTU, desired compression, and desired packet filter identifiers.

2.6.2 Diameter

With the emergence of new wireless technologies and applications, the requirements for AAA have greatly increased, and access control mechanisms have become more complex than ever. Because the existing RADIUS protocol may not be sufficient to cope with these new requirements, a new protocol was defined while keeping the flexibility for further extension. Diameter began to be standardized as a successor to RADIUS in 2000 after the completion of RADIUS standardization. Diameter was designed with a much broader set of requirements in mind; some of them are scalability, transmission level security, certificate transport, mutual authentication, ability to run over IPv4/IPv6, support for proxy and routing brokers, failover, and transferring service-specific attributes. Thus, Diameter is not a brand-new protocol, but is an enhanced version of the RADIUS protocol.

Most of the basic building blocks of Diameter are described in the main specification, Diameter base protocol. It extracts the essence of the AAA protocol from RADIUS and defines a set of core messages. But many of important Diameter applications are defined in separate specifications, except for accounting applications, which are defined in the Diameter base protocol. For example, authentication and authorization in the Diameter NAS application and mobile IP application are specified in companion documents NASREQ [NASREQ] and Mobile IPv4 [DIAMMIP], respectively.

In contrast to RADIUS, which was designed as a client–server protocol for AAA, Diameter is a peer-to-peer protocol that supports more diverse service scenarios and applications. Thus, every node who implements the Diameter protocol can act as either a client or a server depending on network deployment. Communication between Diameter peers starts by one node by sending a message to another node. The Diameter node that receives the user connection request acts as the Diameter client, which is NAS in most cases. The client then receives user credentials, such as username and password, and sends an access request message to a Diameter server. The Diameter server authenticates the user based on the provided information. If the authentication succeeds, the user's access privileges are sent back to the corresponding Diameter client using the response message. Otherwise, an access reject message is sent.

In Diameter, special agent nodes are defined; they are relay, proxy, redirect, and translation agent. A relay agent is for forwarding a message to the appropriate destination and can aggregate requests from different realms (or regions) to a specific realm. Request aggregation can eliminate the burdensome configurations of network access servers for every Diameter server change. A proxy agent can also forward messages, but differently from a relay agent, it can modify the message content for providing value-added services. Through this service, the proxy agent can administrate a specific realm by applying rules on different messages. A redirect agent acts as a centralized configuration repository for other Diameter nodes. When it receives a message, it checks its routing table, and returns a response message along with redirection information to its original sender. With this kind of agent, other nodes may not have to keep a list of routing entries locally, and whenever necessary, they can look up the redirect agent. The translation agent is used for converting a message from one AAA protocol to another. It is useful for integrating the user database of two different application domains, which run their own AAA protocols. It may also be useful in a situation where a company desires a smooth migration to the Diameter protocol, while keeping the backward capability with the current protocol, e.g., the RADIUS protocol.

Since the Diameter protocol is not bound to a specific application, it focuses on general message exchanging features. In the base Diameter protocol, general capabilities negotiation procedures are described. It also defines certain rules that apply to all exchanges of messages between Diameter nodes. This implies that for authentication and authorization, the Diameter base protocol does not define command codes and attribute value pairs (AVPs) (typically is used to encapsulate protocol-specific data, e.g., routing information as well as AAA information) specific to authentication and authorization because the mechanisms may be different from application to application. It is the responsibility of a particular application to determine its own messages and the AVPs to be included in the messages.

2.7 Conclusion

In this chapter, we have given an overview of the basic concepts of information security and various countermeasure technologies. With the emergence of new applications in wireless communication, security concerns have greatly increased recently. This phenomenon is gaining more attention as attack skills are being renovated and new attack techniques are continuously emerging.

Moreover, due to the diverse service scenarios and applications, access control mechanisms and their related technologies have also become ever complex. We expect that the security concerns will be much more amplified with the widespread of smart phones. Smart phones are adept at handling more tasks, including stock trading, paying bills, bank transactions, and on-line shopping. Many experts are warning that as smart phone devices have become more popular, harmful software such as viruses and spyware is emerging to exploit their vulnerability. More seriously, many attacks on smart phones are relying on the legal software that can be downloaded from the application server. A popular one is the mobile phone software that can help someone eavesdrop. Mobile applications, which are sold or distributed through online market, can be exploited as an attractive tool for potential security breaches.

Although the contents described in this chapter are rather brief and not deep for the security professionals, they will help readers remember the security basics while reading more sophisticated issues of security in the next generation wireless communication that are introduced in the rest of this book.

References

[1] M. Nakhjiri and M. Nakhjiri, *AAA and Network Security for Mobile Access*. Wiley, 2005.
[2] C. Boyd and A. Mathuria, *Protocols for Authentication and Key Establishment*. Springer, 2003.
[3] E. Rescorla *et al.*, A survey of authentication mechanisms, Internet draft, Internet Architecture Board, IETF, February 2010.
[4] M. Bishop, *Computer Security: Art and Science*. Addison Wesley, 2003.
[5] B. Schneier, *Applied Cryptography*. Wiley, 1996.
[6] Remote Authentication Dial In User Service (RADIUS), IETF RFC 2865, June 2000.
[7] Diameter Base Protocol, IETF RFC 3588, September 2003.
[8] Extensible Authentication Protocol (EAP), IETF RFC 3748, June 2004.
[9] The EAP Authentication Protocol, IETF RFC 5216, March 2008.
[10] S. Aissi, N. Dabbous, and A. R. Prasad, *Security for Mobile Networks and Platforms*. Artech House, July 2006.

3

Standardization

Technology standardization takes the given solution to global market. Standardization also gives choice to customers to buy products from different vendors that can communicate/work with each other. Thus standardization gives choice to customers, drives cost down while giving global reach to the vendors. There are several standardization bodies worldwide in the field of wireless communications, out of which two are of our concern: Third Generation Partnership Project (3GPP) and WiMAX. In this chapter, we present details of standardization in these two bodies.

3.1 3GPP

The Third Generation Partnership Project (3GPP) is a body for developing technical specifications and reports for mobile communications systems. The original goal of 3GPP was to develop third generation (3G) specification but global system for mobile communication (GSM) specification and its maintenance was also added to it. In this section we briefly present the standardization process of 3GPP and also explain the network architecture of 3GPP as it has evolved.

3.1.1 Organization

As the name suggests, 3GPP is a project among partners where the partners are the organizational partners. These organizational partners are standardization bodies of given country. Thus 3GPP develops specification that is standardized by organizational partners representing various regions of the world. These organizational partners follow their government/regulatory mandate. The organizational partners participate in the Project Coordination Group (PCG) of 3GPP. PCG is the highest decision making body of 3GPP with management, time, and other such responsibilities; it meets once in six months for final adoption of Technical Specification Group (TSG) work items.

There are individual members (vendors, operators etc.) who are members of at least one of the organizational partners who provide input to the TSG. The result of TSG is technical report (TR) or technical specification (TS) that forms (as such or at least basis of) the specification of organizational partners. The 3GPP also takes input from International Telecommunications Union (ITU) and uses its guideline (e.g., for IMT-2000 and IMT-Advanced). The resulting specification from 3GPP TSG is taken to ITU by individual members as specification. This organizational structure in shown in Figure 3.1 [10].

The technical study groups (TSGs) of 3GPP are Radio Access Network (RAN) group, Service and Systems Aspect (SA), and Core Networks and Terminals (CT) with GSM EDGE radio access networks as the fourth one. Each TSG is further divided into working groups (see Figure 3.2 [11]). Security standardization responsibility falls under the SA WG3 (also known as SA3). Each WG has separate meetings on the topics of standardization relevance about four or more times a year.

3.1.2 Standardization Process

To start standardization in a given topic there should be a proposal that is supported by at least four companies, out of which one should be a mobile operator. This proposal for project is called Working Item Description (WID). The WID first goes through an approval in a WG where a consensus should be reached on the given WID proposal even if there are sufficient supporting companies. After a given WID is agreed on by the given WG (e.g., SA3), then it is sent to the plenary meeting of the given TSG for approval (e.g., SA Plenary for SA3 WID because SA3 belongs to TSG SA). A Work Item (WI) can be of the type that requires study and thus is a Study Item (SI) that usually results in a Technical Report (TR), else it could be a mature topic that can be standardized, resulting in a Technical Specification (TS). The numbering of

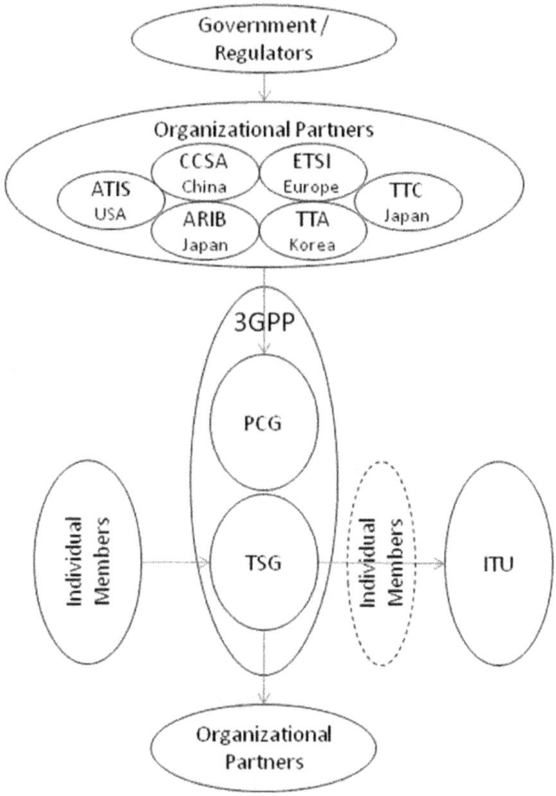

Fig. 3.1 Standardization organization.

TR or TS has a convention of xx.yyy a.b.c; this is explained below [12, 13]:

— "xx" is the type of specification or series, e.g., 33 refers to security or SA3 specification.
— "yyy" is the specification number assigned to a TR or TS. A TR with first y being 8 (e.g., 820) is considered a 3GPP internal TR while one starting with 9 is for publication.
— "a" is the major field.

- If this field is 0 then the given TR or TS is still immature.
- 1 means the work is considered 60% complete and has been presented to a TSG plenary meeting for information.

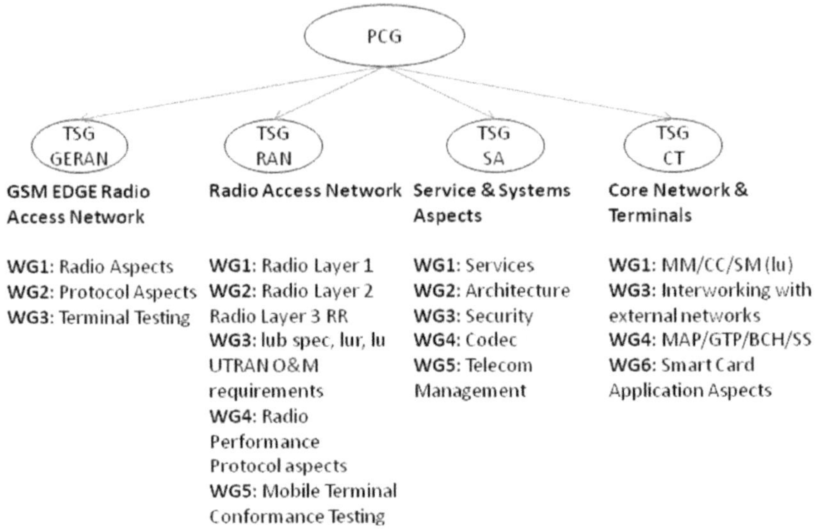

Fig. 3.2 3GPP structure.

  - 2 means that the work is at least 80% complete and is presented at a TSG plenary for approval.
  - 3 or higher number means that the work is approved and is under change control, this number relates to the release of specification the TR or TS belongs to, e.g., a given security specification could belong to Release (Rel.) 6 and thus relates to RAN and architecture of Rel. 6.
    - Release can be seen as the group to which a specification belongs to. Release 99 (or Rel. 9), also known as Rel. 3, was the first one for 3G, after which there was Rel. 00 or Rel. 4 going on to Rel. 10. Now 3GPP is busy with development of Rel. 11 specification. For GSM there were Phase 1, Phase 2 and Phase 2+ specifications.

— “b” is the technical part of the version and is incremented whenever a technical change is made to the TR or TS of a given release.

— “c” is the editorial field that is incremented each time there is a non-technical change in a specification.

Each release of a specification is standardized in three stages:

— Stage 1: refers to services
— Stage 2: gives the functional architecture or solution — security or SA3 specification is usually Stage 2 activity that is then taken by other WGs, e.g., CT1, CT4, RAN2, RAN3 to develop Stage 3 specification
— Stage 3: is concrete implementation details

Different series and their explanation are given in Table 3.1. An easy way to find a specification of a given series is to visit the series website, e.g., for security series 33 visit <http://www.3gpp.org/ftp/Specs/html-info/33-series.htm>, and for a given specification to visit the specification website, e.g., for 33.401 visit <http://www.3gpp.org/ftp/Specs/html-info/33401.htm> where the bottom most is usually the most recent version of the given release of the specification.

3.1.3 Evolution: 3GPP Network and Security

As given in the previous section, the standardization scope of 3GPP is from the subscriber identity module (SIM), to the mobile equipment (ME), RAN,

Table 3.1. Specification series and associated number in 3GPP.

Subject of specification series	3G/GSM R99 and later
Requirements	21 series
Service aspects ("stage 1")	22 series
Technical realization ("stage 2")	23 series
Signaling protocols ("stage 3") — user equipment to network	24 series
Radio aspects	25 series
CODECs	26 series
Data	27 series
Signaling protocols ("stage 3") — (RSS-CN)	28 series
Signaling protocols ("stage 3") — intra-fixed-network	29 series
Program management	30 series
Subscriber Identity Module (SIM/USIM), IC Cards. Test specs.	31 series
OAM&P and Charging	32 series
Access requirements and test specifications	
Security aspects	33 series
UE and (U)SIM test specifications	34 series
Security algorithms (3)	35 series
Evolved UTRA (LTE) aspects	36 series

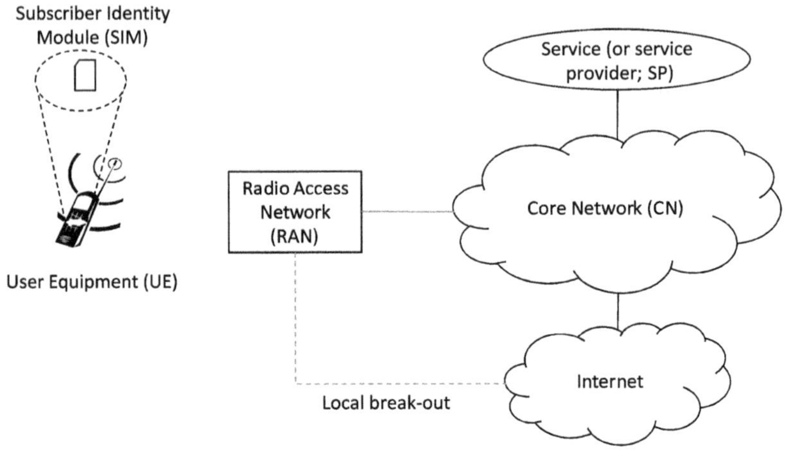

Fig. 3.3 3GPP basic network architecture.

core network (CN), and to some extent services or platform for services like IP Multimedia Subsystem (IMS). Thus the basic network architecture of 3GPP looks as in Figure 3.3. In this section we briefly look at the network evolution in different release. As a background we also briefly discuss GSM security and its issues.

3.1.3.1 3GPP Network Evolution

In this section, we present a brief overview at the 3GPP network evolution in different releases. Figure 3.4 gives an overview of the network evolution from GSM onwards. The figure does not show minor changes in different releases and only major network elements are presented.

Figure 3.4 is divided in four parts: user equipment (UE), access network (AN), core network (CN), and the rest given as services, etc. The UE is a combination of the terminal itself (mobile station (MS) in GSM and mobile equipment (ME) in systems starting from 3G) and the subscriber identity module (SIM). The AN constitutes of the radio part with the base station and controllers of wireless part. The CN takes care of mobility, session control etc. Home Subscriber Subsystem (HSS) with all its elements takes care of subscriber information and authentication.

The basic elements of a 3GPP network are given in Figure 3.4. From evolution of GSM to general packet radio system (GPRS), the CN works

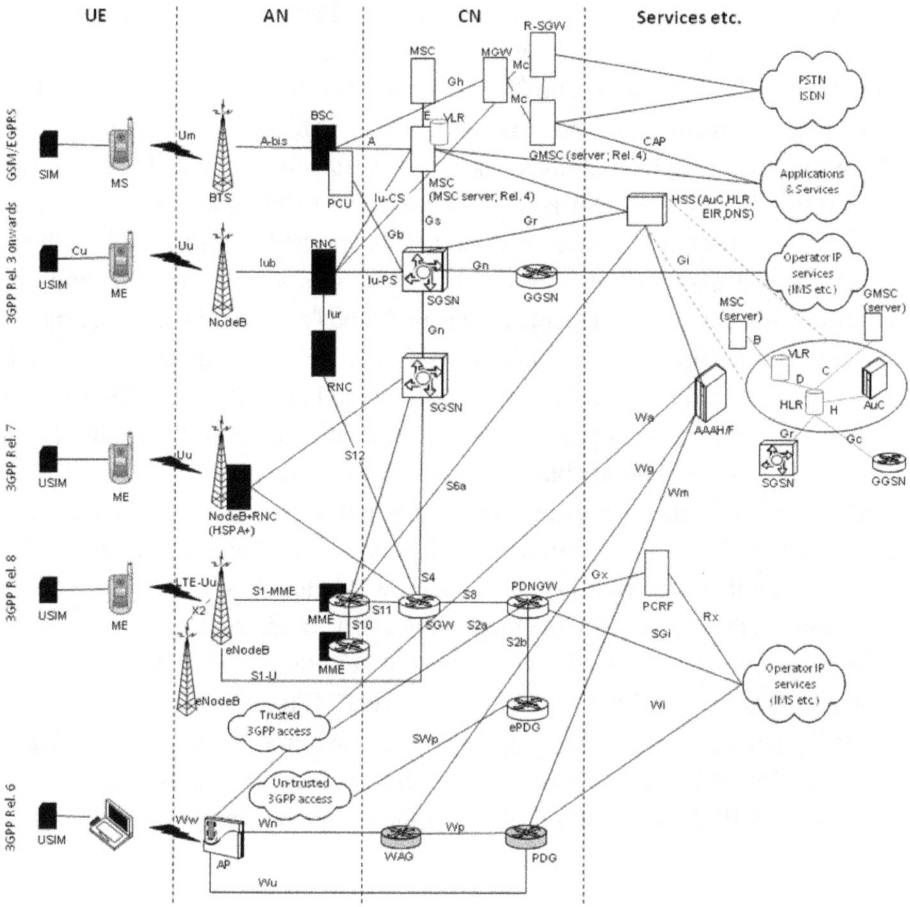

Fig. 3.4 3GPP network evolution.

in packet switched (PS) or circuit switched (CS); more commonly these are known as PS or CS domain. As shown in Figure 3.4, the CS domain is the most legacy one existing from the GSM (or second generation; 2G) era with GPRS or 2.5G bringing the PS domain in mobile networks. In the CS domain the mobile switching centre (MSC), the switching element, is the most important network element usually integrated with the visited location register (VLR) where the VLR contains a database in the location, which belong to the given MSC. Within the PS domain, the two main network elements are the serving GPRS support node (SGSN) and the gateway GPRS support node (GGSN).

Each reference point between two network elements has a protocol stack. Initially the protocol in the CN was signaling system 7 (SS7), which was also valid for Rel. 3 CS domain. The PS domain always used Internet protocol (IP). From Rel. 4 onwards, with the standardization of MSC and GMSC server, the CN started using IP as well as asynchronous transfer mode (ATM) with the HSS using SS7 while ATM was used for interfaces to the AN. From Rel. 5 IuPS was also used for Gb interface between SGSN and BSS and thus all interfaces to AN started using ATM, this is when IMS was standardized. Details of activity/enhancement in each release can be found [14]. Rel. 6 saw the entrance of wireless local area networks (WLANs) in the 3GPP standardization, which was also worked on in Rel. 7. In Rel. 7 a flattened architecture in the form of high-speed packet access plus (HSPA+) came in being with the radio access network (RAN) and the 3G base-station, the NodeB, being combined. Rel. 8 (and beyond) is what we are concerned with in this book; in Rel. 8 3GPP took a bold step of changing the complete AN and CN with interfaces to non-3GPP networks, be it trusted or otherwise. The new AN and CN in Rel. 8 is known as long-term evolution (LTE), which is the evolved universal terrestrial radio access network (E-UTRAN) and the system architecture evolution (SAE) or evolved packet core (EPC), respectively. SAE/LTE together is also known as evolved packet system (EPS). Rel. 8 also saw the introduction of 3GPP femto known as Home NodeB (HNB) for UMTS femto and Home evolved NodeB (HeNB) for EPS based femto. In Rel. 10 the relay node is introduced.

3.1.3.2 3GPP Security Evolution

Together with network evolution there has been enhancement in security for different releases of 3GPP, which is summarized in Table 3.2 [15–21].

3.2 IEEE 802.16 and WiMAX

The IEEE 802.16 standard defines the wireless metropolitan area network (MAN) air interface specification, i.e., the Physical (PHY) and medium access control (MAC) layers. Although the IEEE 802.16 family of standards is officially called *WirelessMAN*, it has been dubbed WiMAX (worldwide interoperability for microwave access) by an industry group called the WiMAX Forum®. The forum started in June 2001 with a single focus to facilitate the deployment

Table 3.2. Security comparison.

	GSM	GPRS	UMTS	SAE/LTE
Security services	Ciphering User authentication by network Made equivalent to wired networks	Ciphering User authentication by network	Ciphering and integrity protection Mutual authentication	Ciphering and integrity protection Mutual authentication
Authentication method	Authentication using 3 values		UMTS-AKA using 5 values	EPS-AKA (similar to UMTS-AKA) using 5 values
Keys	Derivation of a ciphering key after authentication		Derivation of 2 keys one for ciphering and other for integrity protection	Ciphering key for user plane, ciphering and integrity keys for signaling in RAN and CN
Key length	Shared key 128 bits for authentication Derived 64 bits out of which 54 were used for ciphering	Shared key 128 bits for authentication Derived 64 bits for ciphering	128 bits	128 bits
Key handling	Changed on authentication	Changed on authentication	Changed on authentication	Changed on each handover and in other situations
Algorithm	A5/1 (normal)/2 (export)/3. A5 1 and 2 are broken; specification is confidential. A5/3 is based on Kasumi	GPRS Encryption Algorithm (GEA): GEA0(none), GEA1 (export), GEA2 (normal) and GEA3 (A5/3 type)	Kasumi from Rel. 4	SNOW 3G and AES (third algorithm, ZUC, under discussion)
Security end-point	BTS	SGSN	RNC/SGSN	eNodeB for UP and RRC MME for NAS
Network security	None	None in initial release	MAPsec and IPsec based network domain security	IPsec based network domain security

of broadband wireless networks based on the IEEE 802.16 standard wireless access equipments. The mission of the forum is to promote and certify conformity, compatibility, and interoperability of broadband wireless products.

3.2.1 IEEE 802.16 Standards

3.2.1.1 Overview

IEEE 802 refers to a family of IEEE standards dealing with local area networks (LANs) and MANs, which is maintained by the IEEE 802 LAN/MAN Standards Committee (LMSC). The IEEE 802 organization is made up of various types of working groups, each of which provides the focus for each area. Among these, the IEEE 802.16 Working Group, which was established by IEEE Standards Board in 1999, aims to prepare formal specifications for the global deployment of broadband WMAN. The IEEE 802.16 standards specify the structure of the PHY and link layer operations that occur between a subscriber station (SS) and a base station (BS). IEEE 802.16-2001, the first standard of the 802.16 family, was approved in December 2001 and published in 2002. However, it turned out to be an inadequate air interface standard for broadband wireless access, and was soon followed by several amendments. For the amendments, the working group formed several small groups called task group (TG) (e.g., 802.16a or 802.16e), and the TGs have undertaken the amendment projects. While earlier TGs were officially closed after their missions were over, new TGs have succeeded them whenever new requirements appeared. Some of the currently working TGS are 802.16h (License-Exempt TG), 802.16j (Relay TG) and 802.16m (Advanced Air Interface TG).

IEEE 802.16e is an evolved version to support the mobility for WiMAX, thus often called mobile WiMAX. *WiBro* is the South Korean service name for mobile WiMAX, and has been adopted as the sixth global standard for the third-generation telecommunication. More specifically, mobile WiMAX provides enhancements to IEEE 802.16 to support an SS moving at vehicular speeds and thereby specifies a system for combined fixed and mobile broadband wireless access. Mobile WiMAX specifies functions to support an efficient resource allocation scheme in mobile and mesh scenarios as well as higher layer handover between BSs or sectors. Currently, the operation is limited to licensed bands suitable for mobility below 6 GHz. After *WiBro* has been adopted as another global standard for the third-generation

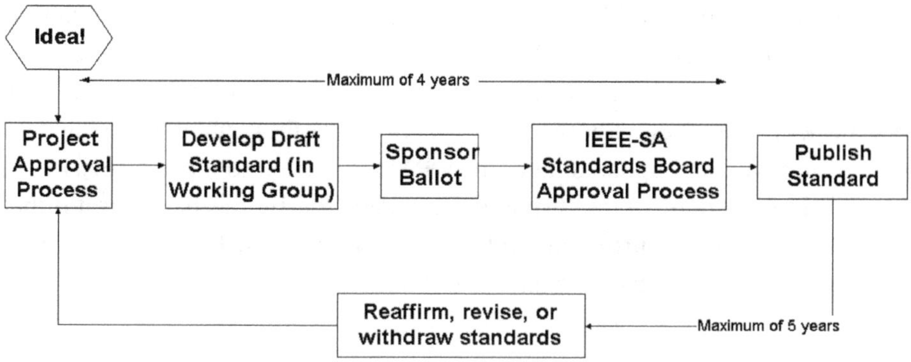

Fig. 3.5 Standards process in a simplified form.[1]

telecommunication, mobile WiMAX is being spotlighted in markets, and consequently, some vendors already included mobile WiMAX-compatible interfaces in their laptops or mobile phones.

3.2.1.2 General standardization procedure of IEEE 802.16

The general development procedure of the IEEE 802.16 standard family used by the Institute of Electrical and Electronics Engineers Standards Association (IEEE-SA) is shown in Figure 3.5.

Figure 3.5 provides a simplified view of the process to develop, approve, and publish a new standard. At the idea phase, an idea for a standard is usually developed by a group of people, and the responsibility for the idea is assumed by the sponsor. The sponsor is usually a society or an existing standards committee. This idea is then transferred onto a form called the project authorization request (PAR) and submitted to the New Standards Committee (NesCom) for approval. The draft is developed and refined by the working group until it is technically complete and the working group decides its readiness for sponsor ballot. The sponsor forms the sponsor ballot group, who votes and comments on the proposed standard. The working group responds to the comments and as necessary produces subsequent drafts for recirculation ballot. After the ballot has achieved consensus, the draft then goes to the Review Committee (RevCom) and the Standards Board for approval.

[1] IEEE 802 Standards Training Courses, http://www.ieee802.org/training.htm.

The life span of a PAR is four years, and an extension can be made if absolutely necessary. The draft is edited and formatted by an IEEE Project Editor and published. The standard is valid for five years before it must be reaffirmed, revised, or withdrawn. A revision allows the content of the standard to be improved and updated, but a new PAR is needed to make a revision to an existing standard. A reaffirmation is a decision by the ballot group that the standard is both current and relevant to today's industry, so does neither require a PAR nor allow the standard to be changed.

3.2.2 WiMAX Forum®

Although necessary, the over-the-air upper layer signaling as well as network architecture and protocols behind the base stations required for an end-to-end specification were considered outside the scope of IEEE 802.16 standards. In addition, 802.16 MAC and PHY specifications, defined with a focus on flexibility, allow many options and various ways of implementations of BS and mobile station (MS) features, which, unless coordinated for interoperability, can result in incompatible products.

The WiMAX Forum was established in 2003 to promote and enable the deployment of WiMAX as a new broadband access technology based on 802.16 standards. To achieve this goal, the WiMAX Forum initiated several technical specification efforts to complement IEEE 802.16 standardization by defining minimum product interoperability requirements or system profiles as well as protocol and radio conformance testing specifications to be used as a basis for industry-wide certification of devices and BSs. In addition to developing testing specifications, the WiMAX Forum has also established certification laboratories and manages the conformance and interoperability testing to ensure that all WiMAX-certified products across different implementations work seamlessly with one another.

The WiMAX Forum is organized in several working groups. The scope of these WGs is given in the Table 3.3.[2]

Application Working Group (AWG) defines *WiMAX™ System Evaluation Methodology*, which captures important aspects of system simulation methodology for a WiMAX™ network, and Service Provider Working Group (SPWG) specifies recommendations and requirements for such networks

[2] WiMAX Forum, http://www.wimaxforum.org/about/wimax-forum-working-groups, March 3, 2009.

Table 3.3. WiMAX Forum working groups and their scope.

Working Group Name	Scope
Application Working Group (AWG)	Define applications over WiMAX that are necessary to meet core competitive offerings and that are uniquely enhanced by WiMAX.
Certification Working Group (CWG)	Handles the operational aspects of the WiMAX Forum Certified program.
Evolutionary Technical Working Group (ETWG)	Maintains existing OFDM profiles, develops additional fixed OFDM profiles, and develops technical specifications for the evolution of the WiMAX Forum's OFDM based networks from fixed to nomadic to portable, to mobile.
Global Roaming Working Group (GRWG)	Assures the availability of global roaming service for WiMAX networks in a timely manner as demanded by the marketplace.
Marketing Working Group (MWG)	Promotes the WiMAX Forum, its brands and the standards which form the basis for worldwide interoperability of BWA systems.
Network Working Group (NWG)	Creates higher level networking specifications for fixed, nomadic, portable and mobile WiMAX systems, beyond what is defined in the scope of 802.16.
Regulatory Working Group (RWG)	Influences worldwide regulatory agencies to promote WiMAX-friendly, globally harmonized spectrum allocations.
Service Provider Working Group (SPWG)	Gives service providers a platform for influencing BWA(Broadband Wireless Access) product and spectrum requirements to ensure that their individual market needs are fulfilled.
Technical Working Group (TWG)	Develop technical product specifications and certification test suites for the air interface based on the OFDMA PHY, complementary to the IEEE 802.16 standards, primarily for the purpose of interoperability and certification of Mobile Stations, Subscriber Stations and Base Stations conforming to the IEEE 802.16 standards.

in *SPWG_Requirements_Release_x.x* from the perspective of network operators intending to deploy WiMAX networks. To extend the scope of interoperability beyond the air interface and based on service providers' requirements, the WiMAX Forum also defines e2e network architectures and protocols specification. Network Working Group (NWG) and Technical Working Group (TWG) define WiMAX Forum Network Architecture Stage 2 and Stage 3, each of which includes *Architecture Tenets, Reference Model and Reference Points* and *Detailed Protocols and Procedures*, respectively. Further information is available at the WiMAX Forum site (http://www.wimaxforum.org/resources/documents/technical/release1).

3.2.3 WiMAX Technology Evolution

The WiMAX technology has continuously been evolved based on the dominant use cases and scenarios from both marketing and technology perspectives. The original 802.16 specification clearly aimed at providing high data-rate,

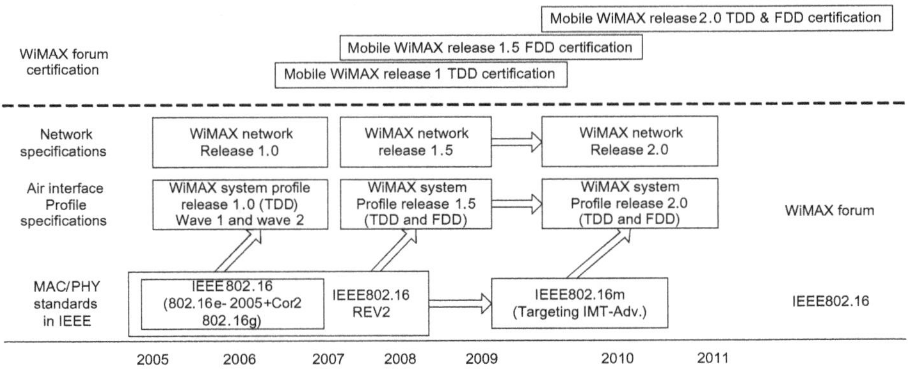

Fig. 3.6 WiMAX technology evolution roadmap.[3]

point-to-point, line-of-sight connectivity between fixed platforms. However, the market needed more diverse requirements such as mobility and point-to-multipoint communication. WiMAX technology development and evolution rely on cooperative and complementary efforts in the WiMAX Forum and IEEE 802.16. Figure 3.6 shows the roadmap of the WiMAX technology.

The main objective of WiMAX System Profile Release 1.0 is to provide OFDMA System Profile specification of a mobile network, complementary to 802.16e-2005 standard, primarily for the purpose of certification of conformant SSs and BS. On the other hand, WiMAX Network Release 1.0, which was developed by WiMAX NWG in 2007, defines the network-level specifications such as the architecture for WiMAX end-to-end network, network discovery/selection, IP addressing, AAA, QoS, mobility supporting, wireless radio resource management, paging, etc.

The work on WiMAX Network Release 1.5, an enhanced network specifications, started in parallel, so that it includes more advanced features such as Robust Header Compression (RoHC), Normative R8, handover data integrity, policy and charging control (PCC), IMS, emergency services (ES), lawful intercept (LI), Simple IP, over-the-air (OTA) provisioning, 3GPP2 inter-working, location-based service (LBS), multicast and broadcast service (MCBCS), universal services interface (USI), Proxy-MIP6, etc.

---

[3]Kamran Etemad, Overview of Mobile WiMAX Technology and Evolution, *Communications Magazine, IEEE*, 46(10): 31–40, 2008.

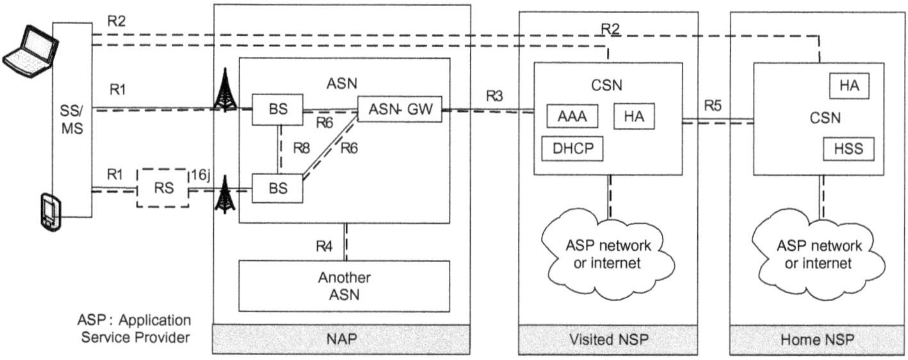

Fig. 3.7 WiMAX network reference model.[4]

A more advanced version, WiMAX network release 2.0 for the next generation WiMAX, will be based on the IEEE 802.16m, which is being developed by the 16m Task Group. IEEE 802.16m is an ongoing project for a new air interface specification intended to comply with the requirements of IMT-Advanced. WiMAX Release 2 mainly aims at enhancing spectrum efficiency, latency, and scalability of the access technology to wider bandwidths in spectrum environments with lots of challenges.

3.2.3.1 IEEE 802.16m and WiMAX Release 2

WiMAX Network Release 2.0 and WiMAX System Profile 2.0 are expected to be based on the IEEE 802.16m, which is targeted for completion in 2010, certification in 2011, and deployment in 2012. The IEEE 802.16m will be a new air interface specification that offers higher spectrum efficiency and throughput to meet the requirements for IMT-Advanced systems. The followings are some of the key enhancements expected in 802.16m:

- Legacy support: Mobile System Profile, Release 1.0
- Lower latency through faster MAC/signaling
- Higher spectrum efficiency through more advanced and higher-order MIMO solutions, including multiuser MIMO as well as lower MAC and PHY overhead

[4]WiMAX Forum, WiMAX Forum® Network Architecture (Stage 2: Architecture Tenets, Reference Model and Reference Points, Part 1).

- Higher peak and user data rates using wider-band carriers (including 20 MHz) and multicarrier aggregation
- Flexible spectrum deployments (both FDD and TDD support in contiguous and noncontiguous bands)
- Improved power saving operation
- Optimizations for improved interworking and coexistence with other access technologies such as 3G systems, WiFi, and Bluetooth
- Support for integrated multihop relay and Femto cells
- Support for government mandates and public safety first responders, military and emergency services such as call-prioritization, pre-emption, push-to-talk.
- Emergency services and communications assistance for Law Enforcement Act.

3.2.4 Security Overview in WiMAX

From the security perspective, the WiMAX system offers relatively stronger security functions. WiMAX security is handled at multiple layers of the network. IEEE802.16 provides a dedicated security sublayer between the MAC layer and physical layer as compared to other existing mobile systems such as cellular systems and Wi-Fi system. The design know-how's are based on the lessons learned from IEEE 802.11 which is known to be highly vulnerable to several kinds of attacks. Besides the security features in IEEE802.16, security in higher layers above the MAC layer of the WiMAX system is defined in the specifications by WiMAX Forum. Each layer of the WiMAX network can be equipped with independent security functions, and some examples of the security mechanisms that can be employed in each layer are shown in Figure 3.8.

The security sublayer in IEEE 802.16 addresses core security functions in the MAC layer. Basically, the security sublayer provides users with privacy, authentication, or confidentiality through applying cryptographic operations to messages carried between SS and BS. That is, MAC layer authentication and authorization ensures that the network is only accessed by permitted users while MAC layer encryption ensures privacy and protects traffic data from eavesdropping by unauthorized third parties. Whereas the IEEE 802.16

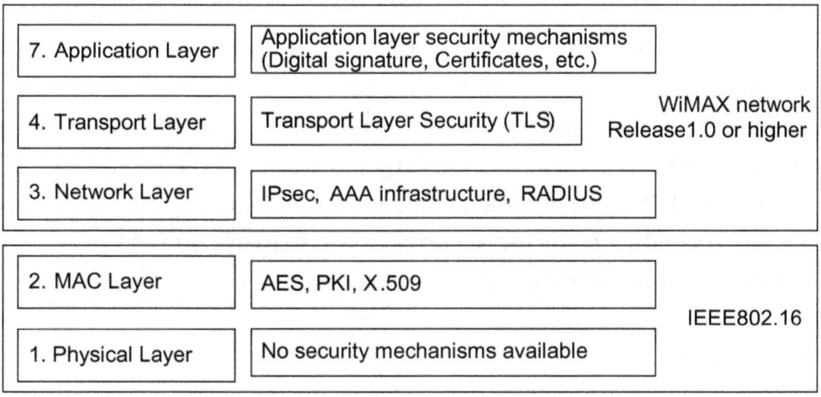

Fig. 3.8 Some possible security mechanisms in each layer of WiMAX.

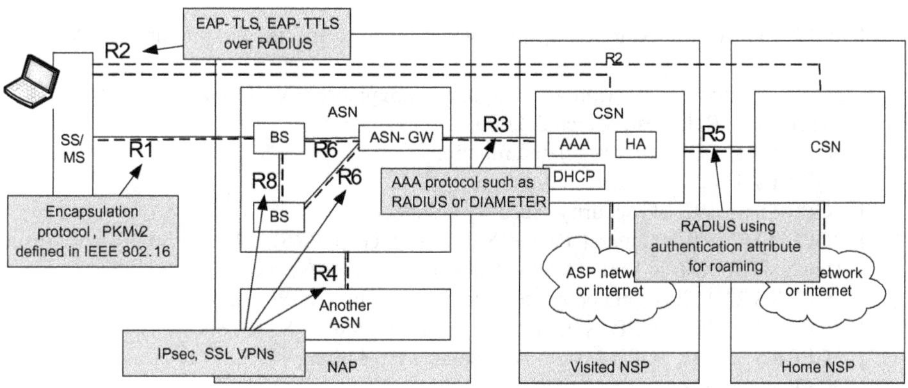

Fig. 3.9 Reference point security in WiMAX.

standards were made for the secure interactions and related protocols between SS and BS, WiMAX network release 1.0 and higher mainly consider the higher layer security, such as network access authentication which is used for authorizing an SS to receive the WiMAX access service and accounting. However, it is noted that the lower layer security and the higher layer security are complementary, and the absence of one is compensated by the presence of the other. Lower layer security between two end points can be a substitute for the higher layer security that terminates on the same end points.

Since the first set of specifications of IEEE802.16 has been released, several types of security vulnerabilities were found to exist. As an effort to fix them,

IEEE802.16 standard group replaced PKMv1 in the earlier version by PKMv2 in the later version IEEE802.16e-2005. However, in spite of the series of amendments in IEEE802.16 and WiMAX Forum specifications, security will continue to be a major concern in the Mobile WiMAX. This is because the Mobile WiMAX system is an IP-based wide area network (WAN) where the device price will be lowered, the resource will be shared by millions of users, and thus a special care for the security of the network and the protection of user information will continuously be necessary.

References

[1] About 3GPP, http://www.3gpp.org/about-3gpp, March 3, 2009.
[2] 3GPP Structure, http://www.3gpp.org/specification-groups, March 3, 2009.
[3] 3GPP Version Numbering, http://www.3gpp.org/Version-Numbering-Scheme, March 3, 2009.
[4] 3GPP Specification Numbering, http://www.3gpp.org/specification-numbering, March 3, 2009.
[5] 3GPP Release Description, http://www.3gpp.org/ftp/Information/WORK_PLAN/Description_Releases/, March 3, 2009.
[6] 3GPP TS 43.020, Security related functions.
[7] 3GPP TS 03.02, Network Architecture.
[8] 3GPP TS 33.102, 3G security; Security architecture.
[9] 3GPP TS 23.401, General Packet Radio Service (GPRS) enhancements for Evolved Universal Terrestrial Radio Access Network (E-UTRAN) access.
[10] 3GPP TS 23.402, Architecture enhancements for non-3GPP accesses.
[11] 3GPP TS 33.401, 3GPP System Architecture Evolution (SAE); Security architecture.
[12] 3GPP TS 33.402, 3GPP System Architecture Evolution (SAE); Security aspects of non-3GPP accesses.
[13] G. Spafford, http://homes.cerias.purdue.edu/∼spaf/quotes.html, March 3, 2009.
[14] ISO/IEC 2382-8, Information technology — Vocabulary — Part 8: Security.
[15] RFC 2828, R. Shirey, Internet security glossary, May 2000, (http://www.faqs.org/rfcs/rfc2828.html).
[16] ISO/IEC 15408-1:2005, Information technology — Security techniques — Evaluation criteria for IT security — Part 1: Introduction and general model.
[17] ISO/IEC 15408-2:2005, Information technology — Security techniques — Evaluation criteria for IT security — Part 2: Security functional requirements.
[18] ISO/IEC 15408-3:2005, Information technology — Security techniques — Evaluation criteria for IT security — Part 3: Security assurance requirements.

4

System Architecture and Long-Term Evolution Security

System architecture evolution and long-term evolution (SAE/LTE), also named as evolved packet system (EPS) and standardized by Third Generation Partnership Project (3GPP) is already seeing worldwide deployment. This is the first time that a mobile system has IP truly coming down to the mobile terminal, i.e., the user equipment (UE) in 3GPP terminology. EPS standardization activity took off with the objectives to come up with a system with higher data rate, lower latency, enhanced quality of service (QoS) and high level of security [TS 22.278]. The objective was also to support variety of access systems while ensuring mobility and service continuity between the access systems. In this chapter we present a brief overview of EPS system and then present EPS security solution in detail.

4.1 Chapter Overview

This chapter covers many topics and therefore in this section we give an explanation on the chapter organization. The chapter is divided in two parts; in the first part (Part I) we discuss the EPS system and in the second part (Part II) we present the details of EPS security.

In Part I of the chapter we start with an overview of EPS system in Section 4.2 together with an overview of the architecture by explaining about different network elements, protocol layers, and different states. After that we present specific points regarding the EPS systems that are relevant for the understanding of the security solution.

Important points regarding the system for better understanding of security are: (1) UE connection to the network that happens in the form of network attachment and also when UE moves to different areas of the network (tracking area [TA]), (2) when the UE is mobile either in the idle state or when the UE performs handover; and (3) identities associated to different aspects of EPS system. A brief description of these points are given below.

- *Network attachment*: This is about UE connecting to the network. We explain how a UE when switched on attaches to the network and give an explanation of security aspects.
- *Tracking area*: A mobile network is divided in different areas where it can be tracked or paged by the network, which is known as tracking area (TA). A very big TA will mean that the network will have to page the given UE over several eNodeBs (eNBs) while small TA means that a mobile UE will have to update and signal its current location often to the network. Change in TA requires signaling between the network and the UE that in turn also means security should be taken in consideration.
- *Mobility*: An EPS system has several types of mobility including handover between and within eNBs, and when the UE is in a given state.
- *Idle state signaling reduction* (ISR): EPS system has ISR functionality that is used to minimize signaling to the core network during idle mode of a UE. Signaling to the core network is known as non-access stratum (NAS) signaling.
- *EPS identities*: EPS has several types of identities for different purposes; we present identities of importance from security aspect.

In Part II, from Section 4.8 onwards, we present security of EPS starting with security requirements. For better understanding of EPS security, we first start with some basic points and then use these basic points for explanation of important security procedures.

We then give an explanation of EPS related keys and EPS security context. Having presented the understanding of the security context, we explain how the security context is handled and an important parameter known as NAS Count. Another basic point to know, and we explain next, is the cryptographic algorithms and their usage. These points explain the basics that we use for detailing the rest of security procedures like authentication etc. The points covered in the security part of the chapter, after security requirements, are briefly explained below.

- Basic points:
    - *Keys and security context*: With the understanding of the EPS architecture the required key types should be obvious. We give an explanation of EPS keys, security context types, and states of these contexts in EPS. There are rules regarding security context handling, e.g. where not to transfer the security context, and specific ways of using security contexts during attachment to the network. Understanding of handling of security contexts is important base for better knowledge of EPS security.
    - *NAS Count*: NAS Count is a value that is incremented for every NAS message sent in the uplink (UE to network) and downlink (network to UE), this value has other important purpose too as explained later in the chapter.
    - Cryptographic algorithms and usage: EPS currently has two sets of integrity and confidentiality protection algorithms and a third algorithm, which is the NULL algorithm. We explain these algorithms and how the algorithms work together with the input parameters required for providing integrity and confidentiality of the messages. We also give an overview of the key derivation function (KDF).
- *Authentication and key agreement*: The basis of security of any system is to identify the communicating parties, which is done by authentication. In EPS the method used for authentication also leads to top level key agreement. The method used is known as EPS authentication and key agreement (AKA).
- *Algorithm negotiation and security activation*: Authentication by itself does not mean that the agreed top level key can be used to derive rest of the keys in the hierarchy and that secure communication can start. The reason being that authentication could be run periodically just for the sake of identifying the communicating

party, which is independent of on-going secure communication. To start using the EPS AKA-based agreed key, a security mode command (SMC) is used. The SMC has the role to negotiate algorithm and activate security in EPS both at the air interface, known as access stratum (AS), and the core network (CN), known as non-access stratum (NAS). Separate SMC is run for NAS and AS.

- *Key handling and state change*: An EPS system can be in several different states, as will be explained in Part I of the chapter; this also means that the security context will be impacted. Thus having explained the security context, authentication and SMC we give an explanation of how the context is dealt with at state changes.
- *Handover key handling*: There are several types of handover in EPS. Handover presents several challenges for security, the crucial one being that keys known by attacker of a specific point must not impact security of future communication. Thus it is necessary to rekey at every handover but then again the rekeying should be done in a very short time so that the on-going service is not impacted. We explain all handover types together with key handling for each handover type.
- *Key-change-on-the-fly*: In EPS, a mechanism is designed for changing keys on the fly even without handover while communication is on-going. This might be necessary for various reasons like the network has new keys/security context that is preferred to be used, etc.
- *Token*: During handover, the Target eNB is prepared with required security context by the Source eNB for secure communication with the UE after handover. Now, there might be cases that handover fails (there could be other cases too) and the UE tries to reconnect elsewhere; in such a case, the message used to reconnect does not have integrity protection thus it is prone to attacks. Therefore in EPS security token is defined that can be used for a sort of local authentication at reconnection without running the full EPS AKA.
- *Local authentication*: This is an authentication feature that is optionally defined in EPS and is based in total number of messages sent.
- *Inter-system mobility*: EPS also provides mobility between other 3GPP systems. As security context is system-dependent, it is

necessary to generate keys during inter-system mobility if ISR is not switched on. Especially for handover case the keys of the target network should be created in a very short interval during which running AKA is not possible. Therefore security context of serving network should be used to create keys for target network, this type of context is known as mapped security context.

- *Voice call continuity* (VCC): As EPS will not provide 100% coverage from day one, it is necessary to provide voice service handover to currently existing circuit switched (CS) systems especially for UE with single radio. We explain the security for such single radio–based VCC (SRVCC).
- *Emergency call security*: Emergency calls should be processed by the network even if the security context of the given UE does not exist.

Part I: SAE/LTE System

Note that SAE/LTE is also known as EPS. In this chapter we use EPS as the terminology.

4.2 EPS System Overview

As explained in Chapter 1, a mobile communications system has a radio access network (RAN) and a core network (CN). Further, in mobile communications systems, there is control plane and user or data plane. The CN of EPS is also known as system architecture evolution (SAE) or evolved packet core (EPC) while the radio access network (RAN) is known as long-term evolution (LTE) or evolved-UMTS terrestrial RAN (E-UTRAN). EPS is the first all IP-based mobile network with IP truly reaching the mobile terminal or user equipment (UE) that is always on and thus always connected. Unlike all previous 3GPP systems, EPS is only packet switched (all IP) thus for voice services it is either dependent on handover to other systems like universal mobile terrestrial system (UMTS) or on IP multimedia subsystem (IMS). In this section, we present the system architecture and the protocol stack of EPS and discuss the functionality of different network elements and protocol layers from security perspective.

4.2.1 Network Elements and Security Functions

EPS architecture is depicted in Figure 4.1 [TS 23.401 and TS 23.402], the interfaces between different network elements (known as reference points and given as S1, X2 etc.) is not explained in this chapter. Dashed lines in Figure 4.1 are for control plane messages and nondashed lines are for user-plane or data

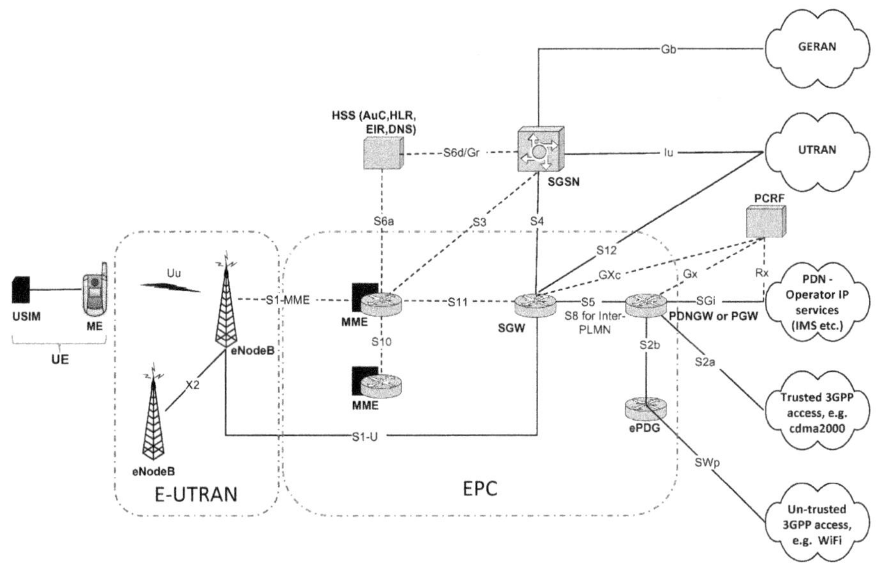

X2, S1-U, S2a, Rx etc. are reference points between network elements. Protocols are defined for each reference point.
Solid lines between network elements are mainly for user plane traffic as defined by 3GPP while dashed lines are mainly for control plane.

AuC	Authentication Center	MME	Mobility Management Entity
DNS	Domain Name System	PCRF	Policy and Charging Rules Function
EIR	Equipment Identity Register	PDN	Packet Data network
EPC	Evolved Packet Core	PDNGW or PGW	Packet Data Network Gateway
E-UTRAN	Evolved-UTRAN	PLMN	Public Land-Mobile Network
ePDG	evolved Packet Data Gateway	SGSN	Serving GPRS Support Node
GERAN	GSM EDGE Radio Access Network	SGW	Serving Gateway
HLR	Home Location Register	UE	User Equipment
HSS	Home Subscriber Subsystem	USIM	Universal Subscriber Identity Module
ME	Mobile Equipment	UTRAN	UMTS Terrestrial Radio Access Network

Fig. 4.1 EPS architecture.

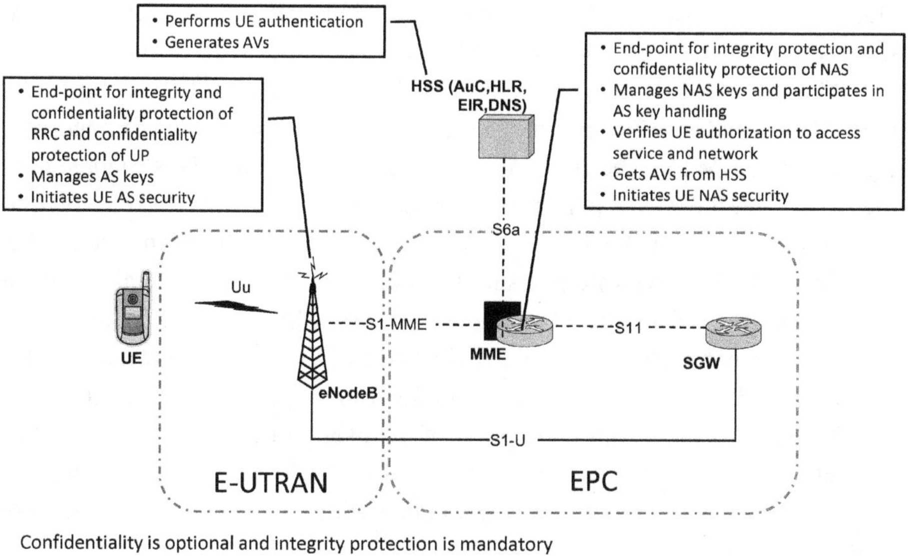

Fig. 4.2 Security at EPS network elements.

traffic. For the sake of simplicity, several details of the architecture are left out and only a basic system description of EPS is given in the following, which is sufficient for better understanding of the security solution. The security functionalities of EPS network elements are depicted in Figure 4.2. Explanation of the network elements is given below.

The eNodeB (eNB) is the radio interface to the mobile network. Integrity and confidentiality protection of access stratum (AS), i.e., the radio resource control (RRC) — the radio level control plane — and the user plane (UP) terminates at the eNB. Thus eNBs also performs AS security-related key handling and algorithm negotiation. Being the radio interface of the network, the eNB takes care of radio resource management (RRM). For efficient usage of air interface, the eNB performs IP header compression. The eNB also has the job of selecting a MME and routing the UP data to the serving gateway (SGW). eNBs also performs handover where intra-eNB does not have to involve the mobility management entity (MME) while inter-eNB handover may involve the MME (see Section 4.13).

The MME provides non-access stratum (NAS) signaling and is the termination point of NAS security. MMEs also perform NAS keys handling, algorithm negotiation, and MMEs participate in key handling of AS security. Connectivity between eNB and MME is secured using IP security (IPsec) [TS 33.210]. MME is involved in handover of a UE within EPS, specifically inter-MME, and also during inter-radio access technology (RAT) handover, e.g., to/from UMTS. The tracking area list (see Section 4.4) is managed by the MME and the MME performs the selection of PDN GW (also known as PGW) and SGW for the UE. The MME interfaces to the home subscriber subsystem (HSS) and participates in AKA of a UE. A MME also takes care of UE reachability during the ECM-IDLE mode (see Sections 4.2.2 and 4.12).

eNBs are connected to a pool MMEs and SGWs. This is known as S1-flex mechanism allowing for redundancy and load sharing in the CN, i.e., the EPC.

The SGW is the termination point of the user plane interface to the eNB; this interface is secured by IPsec. A SGW is the local mobility anchor for inter-eNB handover. For each UE associated with the EPS, at a given point of time, there is a single SGW.

The PDN GW allocates IP address to the UE. It performs user-based packet filtering by, for example, using deep packet inspection (DPI). PDN GW provides accounting for inter-operator charging, packet screening, rate enforcement, etc. A UE can have multiple PDN GWs if accessing multiple packet data networks (PDNs).

A totally flat architecture can be created by combining the SGW, PDN GW and MME [TS 23.401].

4.2.2 Protocol Layers and Security Functions

The mobile network is made of control plane and user plane, the protocol stacks at some of the elements of concern for control and user planes are given in Figure 4.3 and security functionality of relevant ones is shown in Figure 4.4. The protocol layers are discussed below. We assume that the reader of this chapter has an understanding of basics of telecommunications and thus the purpose/functionality of different protocols.

The Layer 1 (L1) or physical layer (PHY) of E-UTRAN is defined in [TS 36.201] and is not of much concern for our discussion here; other layers are discussed below.

The medium access control (MAC) [TS 36.321] provides mapping of logical channels of higher layer to transport channels that are then mapped

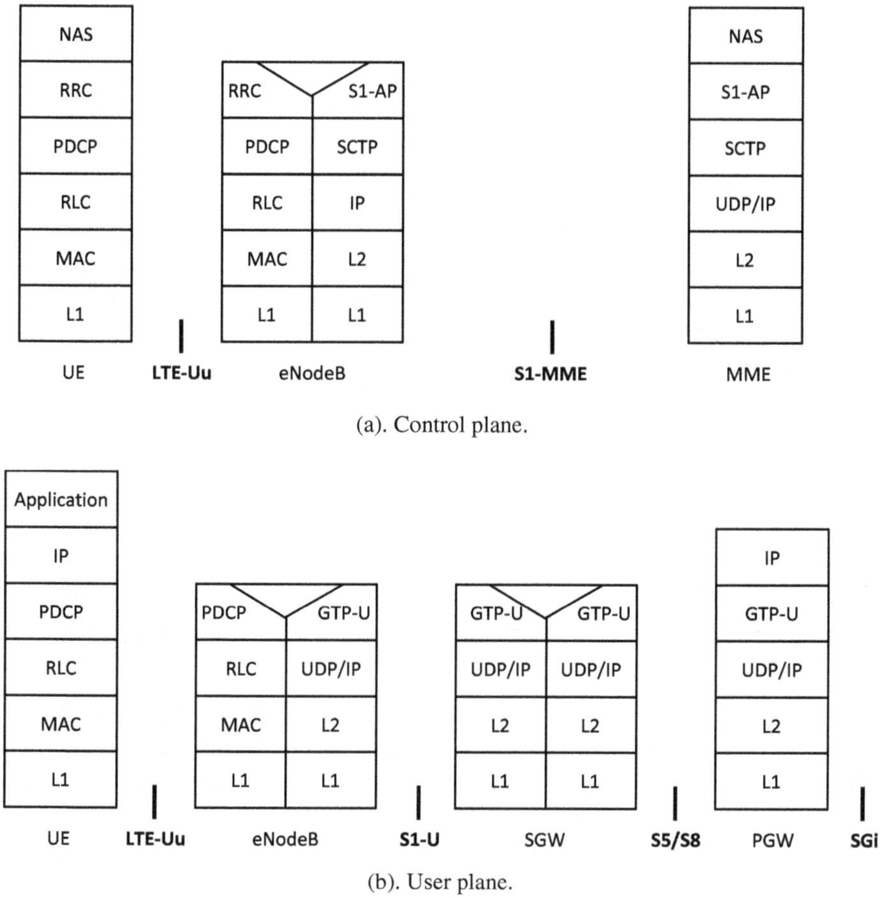

GTP-U	GPRS Tunneling Protocol for User Plane	RLC	Radio Link Control
IP	Internet Protocol	RRC	Radio Resource Control
L1/L2	Layer ½	S1-AP	S1 interface Application Protocol
MAC	Medium Access Control	SCTP	Stream Control Transmission protocol
NAS	Non-Access Stratum	UDP	Universal Datagram Protocol
PDCP	Packet Data Control Protocol		

Fig. 4.3 User-plan and control-plane protocol stacks.

on to the PHY channels. MAC also takes care of priority handling between logical channels of different UEs and priority handling between UEs based on dynamic scheduling. As always, another purpose of MAC is error correction

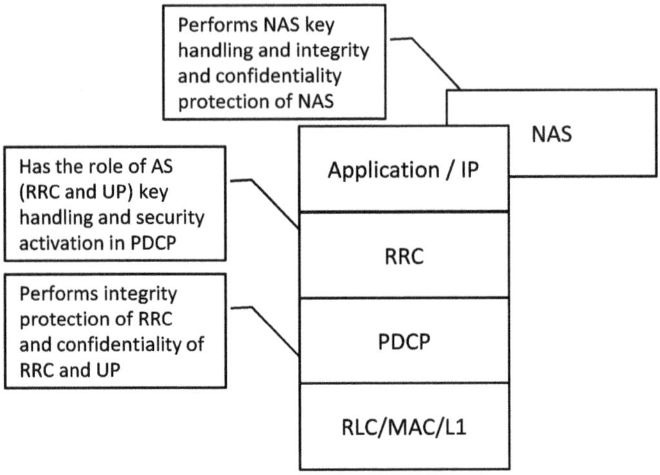

Fig. 4.4 Security and protocol layers relation.

using hybrid automatic repeat request (HARQ). *MAC does not provide any security services.*

Radio link control (RLC) layer takes care of reliable transfer of packets between peers [TS 36.322]. RLC has three different modes out of which one of them is acknowledgement mode for which RLC provides error correction using ARQ. Besides that RLC provides segmentation, assembly, concatenation, in sequence delivery to higher layer and discard of duplicate detected packets. *RLC does not have any security functions.*

The Packet Data Control Protocol (PDCP) [TS 36.323] performs integrity protection of control plane first and then confidentiality protection. PDCP also provides confidentiality for user plane data. For the control plane, the data part of the PDCP packet data unit (PDU) and the message authentication code (MAC-I), used for integrity protection, are confidentiality protected. For the user plane, the data part of the PDCP PDU is confidentiality protected. The data unit that is integrity protected is the PDU header and the data part of the PDU. Radio resource control (RRC) layer activates integrity and confidentiality protection in PDCP and configures the algorithm and key to be used by PDCP. PDCP also provides duplicate detection, retransmission, and in-sequence delivery of packets to higher layer for user plane data. Header compression also happens at PDCP.

The RRC [TS 36.331] layer provides several major functions like connection management, resource release, and connection re-establishment. RRC is the control plane protocol for radio part. Among other things, RRC also performs QoS management, and takes care of mobility and paging. It also provides reliable in-sequence delivery of NAS messages in a cell; NAS messages are either concatenated with RRC messages or carried in RRC without concatenation. For security, RRC takes care of the key management of the Access Stratum (AS), i.e., the user plane and RRC, and security activation. Some RRC messages are not security protected; these are mainly connection setup related messages. RRC can be in two states:

- RRC-IDLE state: This state is reached when the UE is not connected to the network and thus is not known by E-UTRAN. The UE monitors the network for incoming call in this state and also checks network information so as to perform mobility.
- RRC-CONNECTED state: In this state the UE is active and thus user data and signaling can be exchanged between the network and UE.

NAS is the control plane at core network level for (a) mobility management, known as EPS mobility management (EMM), (b) session management, known as EPS session management (ESM), and (c) security management. EMM can be performed after NAS signaling connection is established between the UE and the network. ESM is performed only after EMM procedure has been done and NAS level security is activated except for the first EPS bearer context that is established during EMM's EPS attach procedure. It is in ESM that the IP connectivity for a UE is established and maintained. In terms of security, NAS is the layer where the AKA takes place; it also manages keys for integrity and confidentiality protection of NAS messages. NAS applies integrity and confidentiality to NAS messages independently of the RRC layer but since NAS messages are piggybacked in RRC messages, which are security protected by PDCP, NAS messages are subjected to double security protection. Confidentiality of NAS messages is optional and is an operator's choice. The EMM can be in different states that are related to EMM sublayers of session management and connection management (EPS connection management [ECM]); these states are discussed below and the state model is depicted in Figure 4.5.

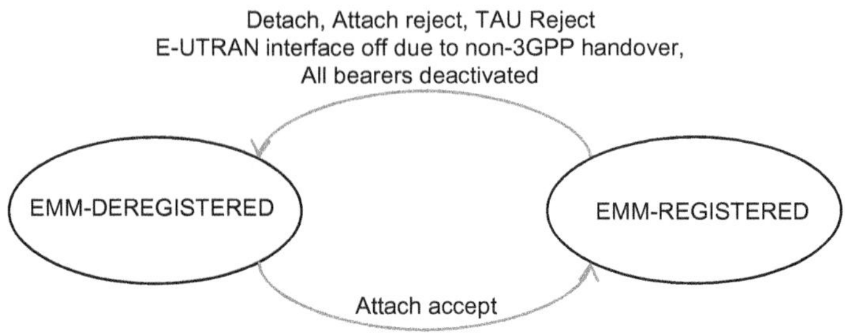

(a) EMM state model in UE

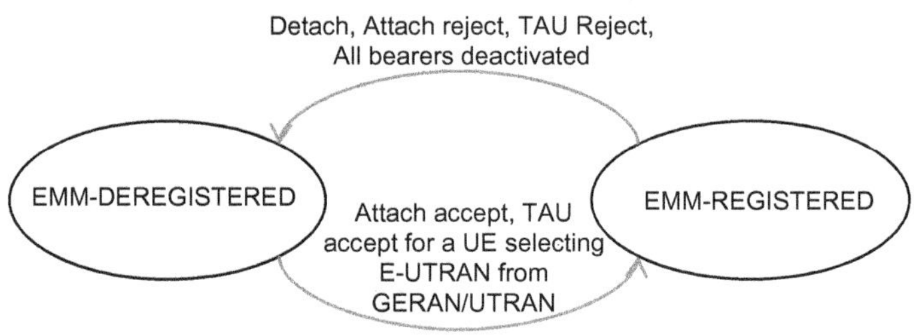

(b) EMM state model in MME

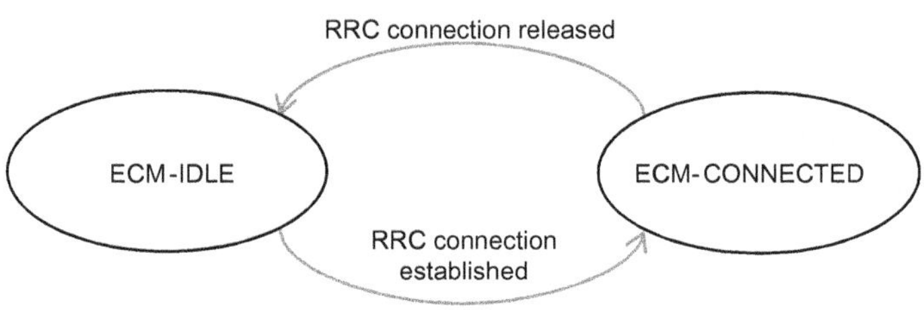

(c) ECM state model in UE

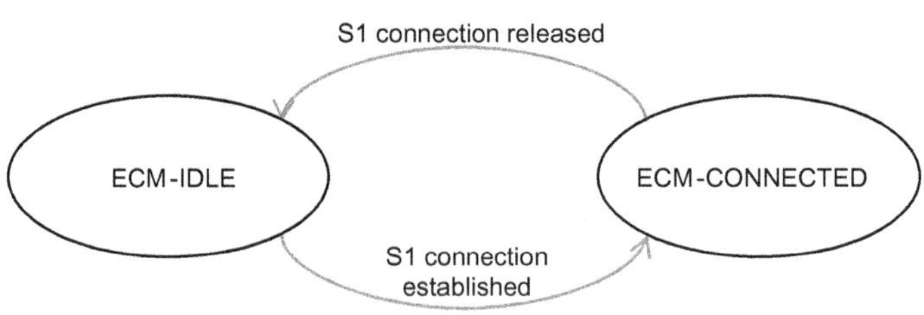

(d) ECM state model in MME

Fig. 4.5 EMM and ECM state models.

- EMM-DEREGISTERED: In this state, the UE is not reachable by the MME because no valid location or routing information about the UE exists. *The security context of UE can be stored in the MME and in the UE.*
- EMM-REGISTERED: In this state, the UE location is known by the MME, the UE has at-least one PDN connection and *the security context is setup*. The UE gets to this sate by attaching to the network while the MME side gets to this state by attaching or when it receives a message from UE informing about its location in the form of tracking area update (TAU) message. The UE can go back to EMM-DEREGISTERED state due to various reasons.
- ECM-IDLE: In this state, there is no NAS level signaling connection between the network and the UE. The UE and MME have security context and the UE is ready to receive paging. The UE can perform mobility and TAU procedure. Procedures like TAU, Attach, Service Request, or detach request take the UE to ECM-CONNECTED state.
- ECM-CONNECTED: The UE is RRC-CONNECTED, EMM-REGISTERED, and ECM-CONNECTED. In this state, the UE can be accessing some service from the PDN thus its location is known by the network at eNB level therefore the UE can perform handover. All security contexts are setup and available in this state. This is the state in which AKA takes place.

In general, the ECM and EMM states are independent of each other. Transition from EMM-REGISTERED to EMM-DEREGISTERED can occur regardless of the ECM state, e.g., by explicit detach signaling in ECM-CONNECTED or by implicit detach locally in the MME during ECM-IDLE. However, there are some relations, e.g., on transition from EMM-DEREGISTERED to EMM-REGISTERED the UE has to be in the ECM-CONNECTED state.

Note that both ECM states and EMM states are related to mobility management thus the TS 24.301 (stage 3 specification) uses EMM-CONNECTED or EMM-IDLE instead of ECM-CONNECTED or ECM-IDLE as is used in TS 23.401 (stage 2 system architecture specification) and TS 33.401 (stage 2 security specification).

4.3 Network Attachment

In this section, we explain the initial network attachment or registration of a UE to the network. A simplified message sequence is depicted in Figure 4.6. Numbered message sequence is explained below.

1. In this sequence, the UE gets access at the radio level by performing random access. The radio level control channel is setup using RRC messages. The RRC connection messages are not security protected. The NAS message called service request can be piggy-backed in RRC message. Service request may or may not

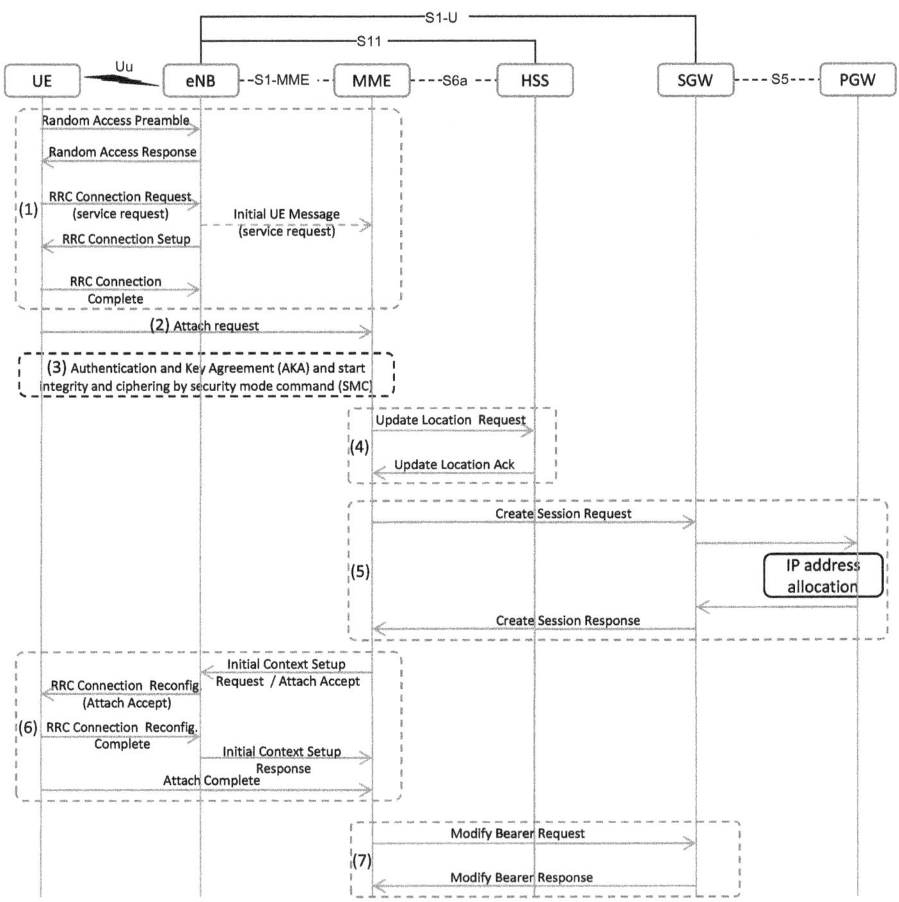

Fig. 4.6 Network attachment, also known as registration.

be security protected, particularly integrity, but in any case, the MME has to act on the message. Before the RRC message is sent the state is RRC-Idle. Service request, a NAS message, changes the state of NAS mobility management (MM) from ECM-IDLE to ECM-CONNECTED.

2. The Attach Request message is sent because the UE needs registration to the network for a service. In this message, all UE information, radio connection-related details and various identities are sent. ME identity can be checked by the network. This message may be security protected if the security context exists but the network will act on it even if it does not have the security context to check the integrity of the message.
3. Here the mutual authentication between the UE and the network is performed known as AKA. After that the NAS and AS level keys are activated by using the security mode commands (SMCs) at NAS level and AS level, respectively. If prior security context exists, then AKA request and response can be already security protected. AKA and SMC are discussed in Sections 4.10 and 4.11 respectively.
4. The HSS is informed at which MME the UE is located.
5. This message sequence is to create bearer between the PDN GW and the eNB. SGW sends MME the contexts to be forwarded to eNB for bearer setup between eNB and SGW. PDN GW is selected by MME based on access point name (APN) from UE or by using default APN setup by the operator. PDN GW assigns the IP address to the UE. The IP address is available at the UE until it detaches from the network thus the UE is always available at the given IP address.
6. This message sequence leads to completion of attach procedure and setting of session parameters. RRC message may be sent without protection but attach accept carried in RRC is integrity protected.
7. Now the setup of bearer between eNB and SGW is completed.

As for network attachment or registration, there is a procedure for detaching from the network. Detach can either be implicit or explicit. Implicit detach can come from network or UE while explicit detach happens when terminal and network have lost connection for a short while. The UE and MME can keep the security context on detach. The MME may get rid of security context after a while in detach, the duration after which the security context is released at the MMEs is implementation-dependent.

4.4 Tracking Area

Tracking area is a part of location management, which itself is part of mobility management in EPS. A MME can manage multiple TAs but an eNB belongs only to one TA. The UE can be connected to multiple TAs, of which the TA identity (TAI) is broadcasted by the eNB. The TAI is constructed of mobile country code (MCC), mobile network code (MNC), and the tracking area code (TAC).

The procedure of tracking area update (TAU) is applied to find the TAI the UE is in, see TAU procedure in Figure 4.7. As given in the figure,

1. the UE sends the TAU Request to the MME,
2. the MME then checks the availability of security context and responds with a TAU Accept or Reject, and
3. if the UE is allocated a new GUTI (a temporary identity, see Section 4.7) then it sends a TAU Complete,

TAU procedure is applied both during handover, i.e., when the UE is active and receiving or sending data, and idle, i.e., the UE is not receiving or sending data; TAU Request is sent if the UE gets in a new TA or when the periodic TAU timer times-out. There are other reasons for running TAU procedure too that we do not cover here; see [TS 23.401 Section on Tracking Area Update

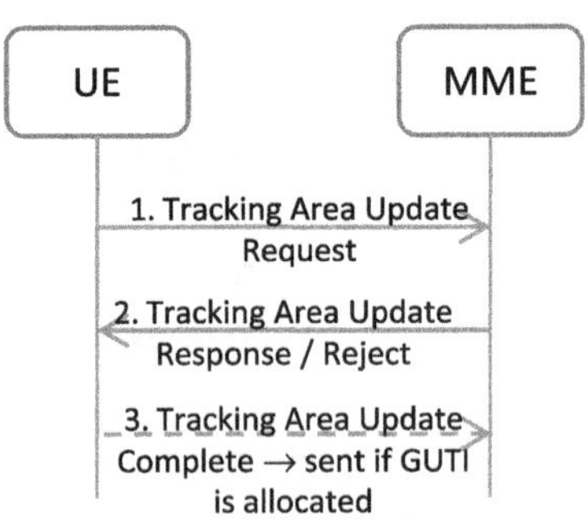

Fig. 4.7 Tracking Area Update (TAU) procedure.

Procedure] for details. A TAI list is provided to the UE by Attach, TAU, or GUTI Re-allocation procedure. Before sending TAU Request the UE must change state from ECM-IDLE to ECM-CONNECTED.

TA dimensioning is very important because too small TA will lead to high volume of messages and if the TA is too large then broadcast messages to a UE will overload all the cells in the TA unnecessarily.

TAU procedure is applied during network attach/initial registration, when the TA is changed, i.e., the TAI is not in the list maintained by the UE, and it can also be applied periodically. Periodic TAU is done for the network to know whether the UE is still available. A change in tracking area can also lead to change in SGW.

The TAU Request (usually just said as TAU) contains the key set identity (KSI) so that the security keys can be identified and matched both at the MME and UE. On receiving TAU Request from the UE, the MME checks for the availability of security keys and rest of the security context so as to perform integrity check of the TAU request. If the MME receiving the TAU request does not have the required keys then it can pull the security keys from other MME or SGSN, the identity of which is provided in the TAU Request. If, after all, the integrity check of TAU Request fails at the MME then a AKA procedure should be run that will in turn also lead to setting of keys for NAS and AS by using SMC. The security aspect of TAU is explained further in Section 4.9.2.3.

4.5 Mobility

There are several types of mobility in EPS that also leads to complicated security procedures. Different types of mobilities possible in EPS are shown in Figure 4.8. Mobility can happen during a session that is known as handover or during idle state known as idle mode mobility. Details of different mobility types are discussed in [TS 23.401, 36.300, 36.331].

As shown in Figure 4.8, basically UE mobility can be within one radio access technology (RAT) or system, which in current discussion is EPS, known as intra-RAT mobility and between different systems or RATs that is known as inter-RAT mobility. There are several types of intra-RAT mobility, such as inter-eNodeB mobility, or mobility between eNodeBs, which can be over the X2 interface (interface between eNBs) or S1 interface (interface between eNB and MME); intra-eNodeB mobility, or mobility between cells of

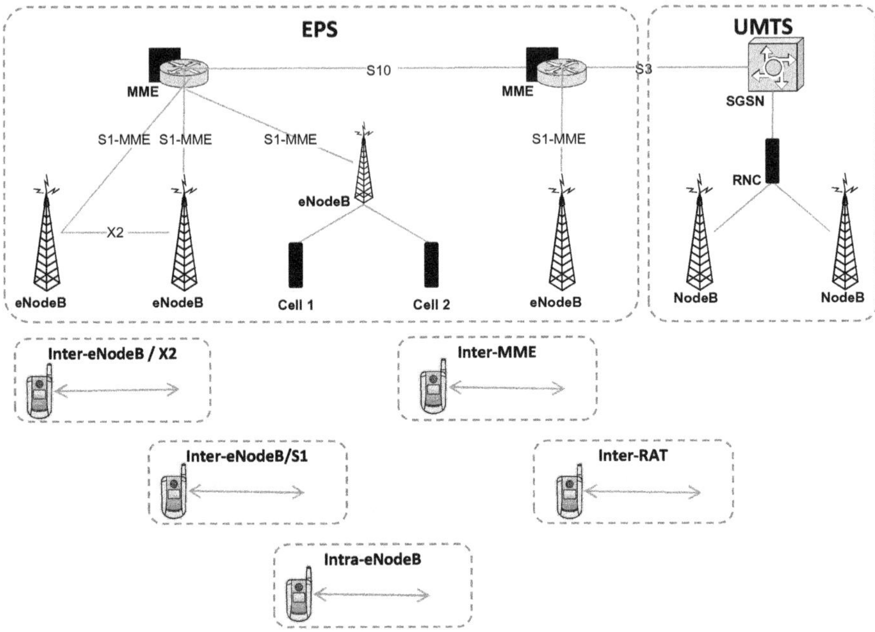

Fig. 4.8 EPS mobility/handover types.

a eNodeB; and inter-MME mobility that happens when the MME is changed during mobility.

A basic handover (HO) message sequence chart for intra-MME handover is shown in Figure 4.9. In this figure the UE performs handover from a Source eNB, where it is currently accessing service, to Target eNB, the eNB where it will access service from. Handover is a RAT decision and the core network is not involved except for a Path Switch message round-trip between the Target eNB, after UE has performed handover to it, and the MME. The Path Switch message is sent during inter-MME handover and might/might not be sent during inter-eNB handover; this message informs MME about the change in eNB for the given UE the MME in turn informs the SGW. The message sequence of Figure 4.9 is explained below.

1. Handover Preparation
   - Based on radio measurement report and radio resource management method the source eNodeB (eNB) decides to perform handover.

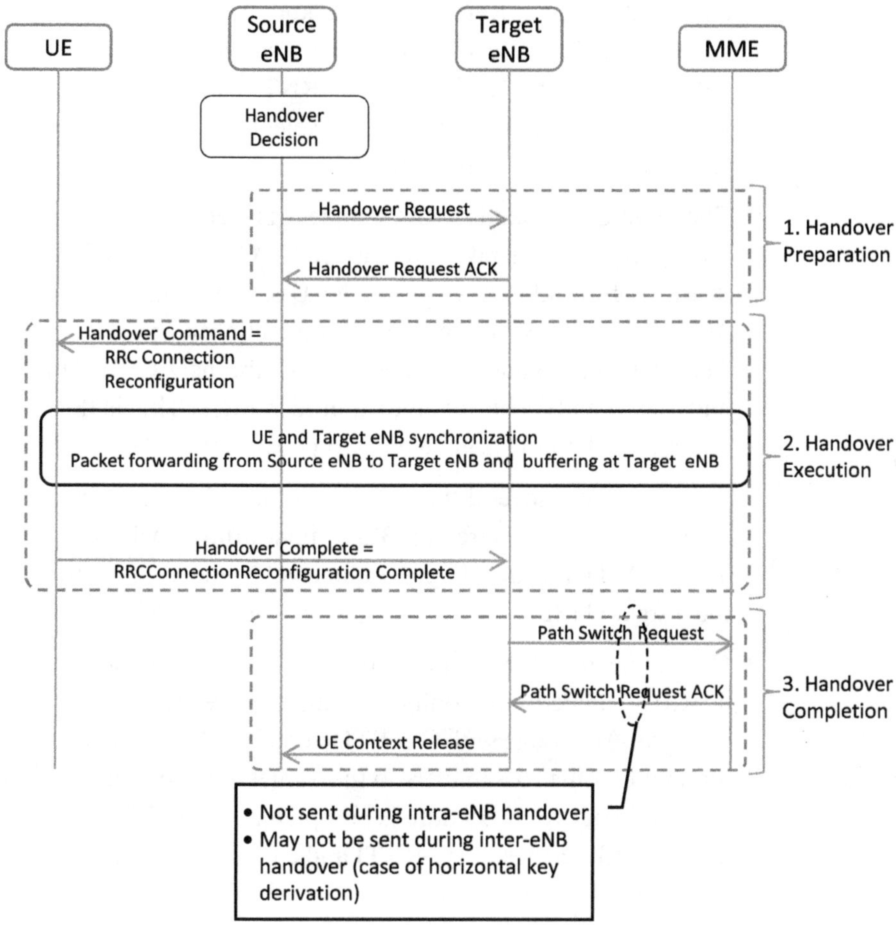

Fig. 4.9 Intra-MME handover showing the handover messages in EPS.

- Source eNB sends a Handover Request message to the Target eNB with necessary information including the so called key K^*_{eNB} (see Section 4.9).

2. Handover Execution

- On receiving the Handover Request, the Target eNB makes necessary preparation at its end including the reservation of C-RNTI — a unique identity for RRC connection — see Section 4.7 on EPS identities for detailed explanation, and sends a Handover Request Acknowledgement message to the Source eNB containing a so called transparent container for UE with a new C-RNTI and Target eNB security algorithm identifiers of the selected security algorithms and other information.

- The Source eNB sends a RRCConnectionReconfiguration (Handover Command, given as RRC Conn. Reconfig. in the Figure 4.9) message to the UE. This message is integrity protected and may be confidentiality protected. The message contains parameters like the new C-RNTI and Target eNB security algorithm identifiers. This is the commanded by the Source eNB to perform the HO.
- As a next step, several things happen including buffering of packets at the Source eNB and forwarding packets to Target eNB while the UE gets radio level connectivity with the Target eNB.
- On connectivity with the Target eNB, the UE sends a RRCConnectionReconfigurationComplete (Handover Complete) message with C-RNTI to confirm the handover, along with other parameters to the Target eNB to indicate that the handover procedure is completed for the UE. The Target eNB verifies the C-RNTI and can now begin sending data to the UE.

3. Handover Completion
    - The Target eNB sends a Path Switch message to MME to inform that the UE has changed cell. The SGW is informed about the handover and takes appropriate action.
    - The MME confirms with the Path Switch Acknowledge message.
    - Target eNB sends a UE Context Release message to the Source eNB the triggers the release of resources by the Source eNB.

4.6 Idle State Signaling Reduction

The purpose of idle state signalling reduction (ISR) is to reduce the frequency of TAU and routing area update (RAU) procedures caused by UEs reselecting between E-UTRAN and UTRAN, both networks operated by one mobile network operator (RAU is in essence similar to TAU but is used in UMTS thus in UTRAN). ISR support is mandatory for E-UTRAN UEs that support GERAN and/or UTRAN and optional for the network. ISR requires special functionality in both the UE and the network (i.e., in the SGSN, MME, Serving GW and HSS) to activate ISR for a UE. The network can decide ISR activation individually for each UE. ISR is enabled only when the UE is able to register via E-UTRAN and via GERAN/UTRAN. Once ISR is activated, it remains active until the TAU Accept message does not contain ISR Active indication. When ISR is activated, this means the UE is registered with both MME and SGSN, i.e., E-UTRAN and UTRAN. In idle state, the UE can reselect between E-UTRAN and GERAN/UTRAN within the registered RA and TAs without any need to perform TAU or RAU procedures with the network. Thus during ISR the UE has security context of both EPS and UMTS (GERAN/UTRAN) because it has performed AKA in both the networks.

4.7 EPS Identities

There are several identities involved in EPS [TS 23.401, TS 23.003, TS 36.300]. It is important to have some level of understanding of these identities to understand the security of EPS. Therefore some of the identities of concern are explained in this section (see below):

- International Mobile station Equipment Identity (IMEI): As the name suggests, this is the identity of the equipment or UE. As such IMEI should be unique for each UE but in reality there are several cases of reuse of IMEI by manufacturers. The IMEI is composed of (1) the type allocation code (TAC) of 8 digits, which is allocated by a central body to all manufacturers; (2) A serial number (SNR) of 6 digits uniquely identifying a mobile terminal within a TAC and allocated in sequential order by a manufacturer; (3) A spare digit which is zero if transmitted.

- International Mobile station Equipment Identity and Software Version Number (IMEISV): The IMEISV is composed of TAC, SNR and a software version number (SVN) of 2 digits to identify the software version of the mobile terminal.
- International mobile subscriber identity (IMSI): This is the unique identity associated with the subscriber. IMSI should not exceed 15 digits and is composed of (1) MCC consisting of three digits, MCC identifies uniquely the country of domicile of the mobile subscriber. The MCC is administered by ITU-T and allocated according to ITU-T recommendation; (2) MNC consisting of two or three digits — this identifies the home PLMN of the mobile subscriber. The length of the MNC (two or three digits) depends on the value of the MCC. (3) Mobile subscriber identification number (MSIN) identifying the mobile subscriber within a PLMN. The MNC and MSIN together form the national mobile subscriber identity (NMSI); this is assigned by national authority.
- Temporary mobile subscriber identities (TMSI) is allocated to mobile subscribers by VLRs, SGSNs, and MME in order to support the subscriber identity confidentiality. The VLR, SGSN, and MME must be capable of correlating an allocated TMSI with the IMSI of the MS to which it is allocated. An MS may be allocated three TMSIs, one for services provided through the MSC this is the TMSI, one for services provided through the SGSN, the packet-TMSI (P-TMSI) and one for the services provided via the MME, i.e., the MME-TMSI or M-TMSI that is part of GUTI.
- Globally unique temporary identity (GUTI) [TS 23.003]: The MME allocates a GUTI to the UE, the purpose of which is to provide identification of UE without revealing its permanent identity. GUTI also allows identification of the MME that allocated the GUTI to the UE. The GUTI is constructed of the globally unique MME Identifier (GUMMEI) and the M-TMSI. Where the GUMMEI is constructed of the Mobile Country Code (MCC), Mobile Network Code (MNC) and MME Identifier (MMEI); while the MMEI is constructed of MME Group ID (MMEGI) and MME Code (MMEC). The UE is identified by the M-TMSI in the MME. A shortened form of GUTI the S-temporary mobile subscriber

identity (S-TMSI) is used for efficient radio signaling procedures (e.g., paging). The S-TMSI is constructed of the MMEC and the M-TMSI.

- Temporary identity used in next update (TIN): As the name suggests, TIN can take the value of any of the temporary identifiers like GUTI, P-TMSI, and TMSI.
- eKSI or key set identifier in E-UTRAN may be either of type KSI_{ASME} or of the type KSI_{SGSN}. A eKSI will be stored in the UE and the MME together with K_{ASME} and the temporary identifier GUTI, if available.
- TAI: This is the identity used to identify tracking areas. The tracking area identity is constructed from the MCC (mobile country code), MNC (mobile network code), and TAC (tracking area code).
- C-RNTI or Cell-Radio Network Temporary Identifier is the unique identification used for identifying RRC Connection and scheduling.

Part II: SAE/LTE Security

4.8 Security Requirements

EPS security is based on the following basic requirements:

- Continued usage of current USIM, i.e., there should not be any change in USIM for accessing EPS network. The USIM that is used in UMTS networks should be thus reusable.
- Security should be at least of the same level or better than that compared to UMTS.

A big difference in EPS compared to UMTS is that the eNB has much more functionalities and it is the location where the AS security terminates. Being the AS security end-point, the eNB stores keys for the AS security and therefore security requirements on eNB is much higher than that for UMTS's NB. Another difference of EPS to UMTS is that the AS and NAS have different termination point therefore they have different requirements. Further, separate key should be used for integrity and confidentiality.

In case of EPS the AS is made of RRC and UP where RRC provides mandatory integrity protection and optional confidentiality protection while

UP provides optional confidentiality protection and no integrity protection. NAS, on the otherhand, provides mandatory integrity protection and optional confidentiality protection. For UP integrity protection is not required because the received packet might be erroneous as probability of error is much higher in the wireless interface. The integrity check of received erroneous packet will fail and thus the packet will not be delivered to the application layer that will lead to negative quality of experience (QoE) impact on the subscriber and probably failure of the service. Further, it makes no sense to repeat packets of real-time services by error correction scheme while any attack on UP — someone sending packets on behalf of any the communicating side — will be easily identified.

The security requirements of EPS are depicted in Figure 4.10. Several important points related to EPS security requirements are discussed above. In case of UE, the requirement is that the MSIN and IMEI(SV) should be confidentiality protected and thus sent only after the NAS security is activated.

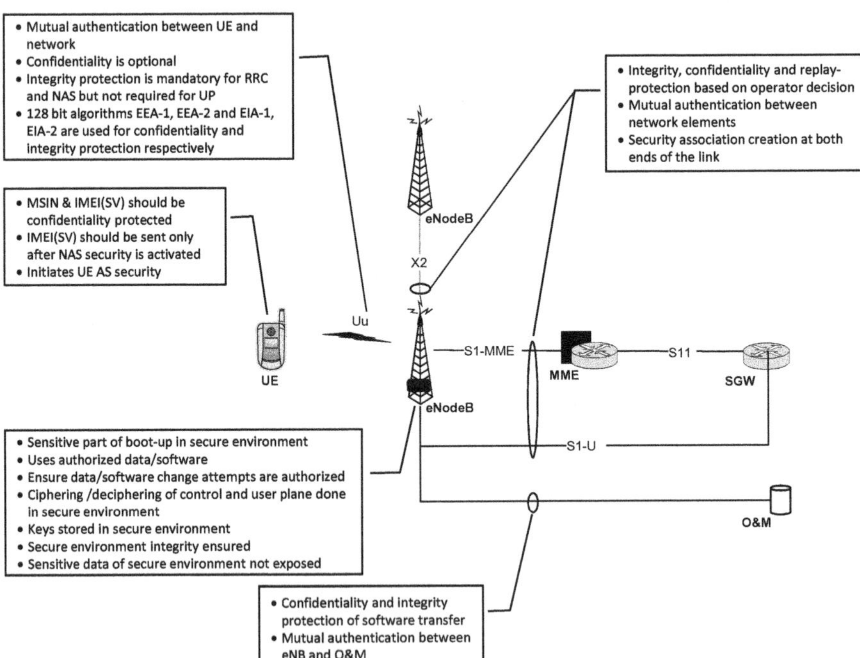

Fig. 4.10 EPS security requirements.

The UE and network should be authenticated to each other, mutual authentication, or more precisely the USIM and the network.

The interfaces X2 and S1 must provision integrity, confidentiality, and replay protection both for user traffic (X2-U and S1-U) and control plane (X2-C and S1-MME). Now an operator can say that the interfaces are secured by some physical means that fulfill the given requirements, then cryptographic means are not required. The cryptographic means to provision integrity, confidentiality, and replay protection for X2 and S1 are based on IPsec ESP tunnel mode or optionally by transport mode [TS 33.401, RFC 4303, TS 33.210 and TS 33.310]. Similarly security is provided for management interface, S1-M, between eNB and operations and management (O&M). Of course, mutual authentication is also done between the associated network elements over the given interfaces.

4.9 Basic Points

In this section, we discuss some of the important security-related basic points. We start the section with an explanation of the keys used in EPS. These keys are part of security context. Therefore next we explain the security context and the states these contexts can be in.

4.9.1 Key Hierarchy and Key Lifetime

EPS has a new key hierarchy, compared to UMTS, due to different termination points for AS and NAS messages. Further, the purpose for AS (RRC and U-plane) messages and NAS messages are different, which requires different keys depending on the purpose of the message because attack on a given message type should not lead to attack on other type of message. Background discussion on key separation for different purpose is also given in Section 4.8.

In addition to the above, eNB is vulnerable to attacks thus keys of a UE compromised at a given eNB should not lead to future threat to UE communication after handover to other eNB. Therefore forward security solution is necessary that is achieved by rekeying at every handover. Rekeying at every handover sets a very high requirement on the time for key generation. Long delay in rekeying during handover will lead to service interruption thus the obvious choice of running complete authentication is not possible as it is time-consuming. Therefore a fast rekeying solution is needed.

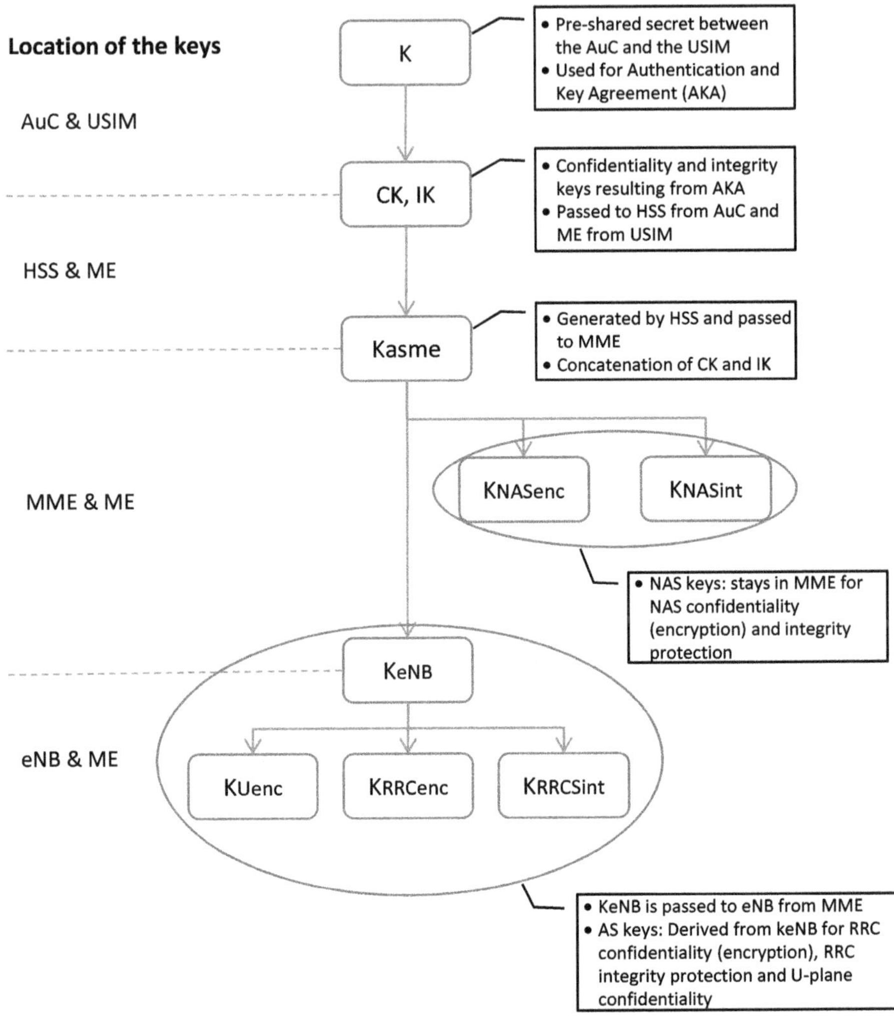

Fig. 4.11 Key hierarchy.

With all the above reasons a separate key for each purpose is required which leads to a key hierarchy as given in Figure 4.11. An explanation of the keys is given below; all key derivation is done by using the key derivation function (KDF) given in Section 4.9.4.5:

- K: This is a secret key of 128 bits that is shared only between the USIM and the authentication center (AuC) of the users home environment (HE), i.e., the operator where the user is subscribed. The key is used for mutual authentication between the network and the UE using AKA. There is no rekeying of this key, i.e., this key has a lifetime as long as the subscription itself.

- CK and IK: The confidentiality and integrity protection keys resulting from AKA that is used for UMTS. Both CK and IK are 128 bits long. At the UE these keys are generated in the USIM and then passed on to the ME while in the network the key is generated in the AuC and then passed to the home subscriber subsystem (HSS).
- K_{ASME}: This key, 256 bits long, forms the root for rest of the keys in EPS. It is derived in the ME and HSS using a key derivation function (KDF) (see Section 4.9.4.5) with input parameters being CK, IK, and serving network identity (SNID). Exclusive OR (XOR) of sequence (SEQ) and anonymity key (AK) from EPS AKA as part of authentication token (AUTN) can also be input parameter. At the network side the K_{ASME} is sent to the MME as part of EPS AKA procedure. K_{ASME} and rest of the keys derived from it are identified by the key set identifier (KSI_{ASME}), that is, a type for eKSI (another type is KSI_{SGSN}). The HSS ensures that K_{ASME} is only sent to the network, of which the SNID was used to derive the key and K_{ASME} is never transferred to an entity out of EPC. Each K_{ASME} is associated with its uplink and downlink NAS Count that is incremented for each NAS message and is never reset. K_{ASME} together with uplink and downlink NAS Count and UE capabilities are stored in the USIM or non-volatile memory (NVM) of the ME.

CK, IK, and K_{ASME} are generated during EPS AKA and when inter-system (or inter-RAT (radio access technology)) mobility takes place, i.e., mapped context is created (see Section 4.17). EPS AKA and thus rekeying of these keys is on operator's configuration. Various reasons that lead to AKA are given in Section 4.10. AKA also takes place when NAS Downlink/Uplink Count is coming to an end, i.e., will wrap around, because repetition of NAS Count with the same key will lead to key-stream reuse.

- K_{NASenc}: This key is used solely for NAS message confidentiality protection using a chosen algorithm (see algorithm types in Section 4.9.4.3). It is derived from the identity of the confidentiality protection algorithm and K_{ASME} using the KDF. The key is 128 bits long and resides in the MME and the ME.

- K_{NASint}: This key is used solely for NAS message integrity protection using a given algorithm; see algorithm types in Section 4.9.4.4. It is derived from the identity of the confidentiality protection algorithm and K_{ASME} using the KDF. The key is 128 bits long and resides in the MME and the ME.

NAS keys do not change until the algorithm to be used is changed, which usually would happen during inter-MME (S1) handover to a MME with different algorithms as higher priority. Input parameters to the KDF for derivation of NAS keys are the algorithm identity (see Section 4.9.4) and the K_{ASME}.

- K_{eNB}: This is a 256 bits key that is initially derived after AKA from K_{ASME} and NAS Uplink Count in MME and ME. In the network side, the MME sends the K_{eNB} to eNB. K_{eNB} is used to derive other AS keys (K_{RRCenc}, K_{RRCint}, and K_{Uenc}) as discussed below. The NAS Uplink Count is a counter that is incremented for each NAS message sent from the UE to the MME. K_{eNB} derivation depends on handover type and changes at each handover, thus also the keys derived from it. Input parameters for the derivation of the initial K_{eNB} are NAS Uplink Count (see Section 4.9.3) and K_{ASME}. After handover the K_{eNB} is equal to the K^{*}_{eNB}.
- K^{*}_{eNB}: This key is derived at each handover using the Next Hop (NH) parameter or K_{eNB} together with Physical Cell ID (PCI) and E-UTRA absolute radio frequency channel number-down link (EARFCN-DL) as input to the KDF. This key is passed to the Target eNB where it is used as the K_{eNB}. See Section 4.13 for details.
- K_{RRCenc}: This key is used solely for RRC message confidentiality protection using a chosen algorithm. It is derived from the identity of the confidentiality protection algorithm and K_{eNB} using the KDF. The key is 128 bits long and resides in the eNB and the ME.
- K_{RRCint}: This key is used solely for RRC message integrity protection using a given algorithm. It is derived from the identity of the confidentiality protection algorithm and K_{eNB} using the KDF. The key is 128 bits long and resides in the eNB and the ME.

- K_{Uenc}: This key is used solely for U-plane confidentiality protection. The key is derived from the identity of the confidentiality protection algorithm and K_{eNB} using the KDF. The key is 128 bits long and resides in the eNB and the ME.

K_{eNB} is derived whenever a decision is taken to start AS security; there are states when NAS keys are sufficient (see Section 4.12). Activation of AS security means that all AS keys are derived. At every handover a K^*_{eNB} is created which is transformed to K_{eNB} (see Section 4.13), out of which other AS keys are also generated.

4.9.2 Security Contexts

4.9.2.1 Security context types and states

In EPS, the security context can be in different states and of different types. The security context in EPS includes keys, cryptographic algorithm chosen and uplink and downlink NAS Count values. An EPS security context has type "mapped," "full native," or "partial native." Its state can either be "current" or "non-current." A context can be of one type only and it can be in one state at a given time. The state of a particular context type can change over time. In terms of context type, only a partial native context type can be transformed into a full native context type. No other type transformations are possible. Definition and explanation of states and types of security context is given below while security context handling is discussed in section below.

Some definitions of contexts are given below for better understanding of states and key types.

- *EPS security context*: A state that is established locally at the UE and a serving network domain. At both ends "EPS security context data" is stored, which consists of the EPS NAS security context, and the EPS AS security context.
- *EPS NAS security context*: This context consists of K_{ASME} with the associated key set identifier, the UE security capabilities, and the uplink and downlink NAS Count values. In particular, separate pairs of NAS Count values are used for each EPS NAS security contexts, respectively. The distinction between native and mapped

EPS security contexts also applies to EPS NAS security contexts. The EPS NAS security context is called "full" if it additionally contains the keys K_{NASint} and K_{NASenc} and the identifiers of the selected NAS integrity and confidentiality algorithms.

- *EPS AS security context*: The cryptographic keys at AS level with their identifiers, the Next Hop (NH) parameter, the Next Hop Chaining Counter (NCC) parameter used for next hop access key derivation, the identifiers of the selected AS level cryptographic algorithms, and counters used for replay protection. Note that the EPS AS security context only exists when cryptographically protected radio bearers are established and is otherwise void. NH and NCC need to be stored also at the MME during connected state.

The context types are as follows:

- *Mapped security context*: Security context created by converting the current security context in MME to a security context for the target system in inter-system mobility, e.g., UMTS keys created from EPS keys. The EPS NAS security context of a mapped security context is full and current.
- *Native EPS security context*: An EPS security context whose K_{ASME} was created by a run of EPS AKA. There are two types of native EPS security context. These are as follows:
    - *Full native EPS security context*: A native EPS security context for which the EPS NAS security context is full according to the above definition. A full native EPS context is either in the state "current" or "non-current"; see explanation below.
    - *Partial native EPS security context*: A partial native EPS security context consists of K_{ASME} with the associated key set identifier, the UE security capabilities, and the uplink and downlink NAS Count values, which are initially set to zero before the first NAS SMC procedure for this security context. A partial native EPS security context is created by an EPS AKA, for which no corresponding successful

NAS SMC has been run. A partial native context is always in state "non-current".

The keys can be in two states as given below:

- *Current EPS security context*: The security context that has been activated most recently. Note that a current EPS security context originating from either a mapped or native EPS security context may exist simultaneously with a native non-current EPS security context.
- *Non-current EPS security context*: A native EPS security context that is not the current one. A non-current EPS security context may be stored along with a current EPS security context in the UE and the MME. A non-current EPS security context does not contain an EPS AS security context. A non-current EPS security context is either of type "full native" or of type "partial native."

4.9.2.2 Security contexts handling general rules

In this section, we present some points regarding handling of EPS security context. The Next Hope (NH) parameter and Next hop Chaining Counter (NCC) are discussed in Section 4.13.

As a basic rule, the current native EPS security context is stored in the USIM or NVM of the ME if there is a change in state to EMM-DEREGISTERED. The stored values are K_{ASME} and NAS Uplink and Downlink Count values. Details with reasons for state change and associated behavior is given in Section 4.12. Another point is that, whenever a key from the USIM or NVM is used, then the stored value is marked as invalid so that it cannot be used once again and when the values are put in the USIM and NVM, then they are marked as valid. The rules for deleting security context from the ME are as follows:

- The UICC is removed from the ME when the ME is in power on state.
- The ME is powered up and the ME discovers that a UICC different from the one which was used to create the EPS security context has been inserted in the ME.

- The ME is powered up and the ME discovers that no UICC has been inserted in the ME.

Other rules of security context handling are as follows:

- The K_{ASME} is never transferred from the EPC to an entity outside the EPC.
- Both the ME and MME are capable of storing one non-current EPS security context and one current EPS security context in the volatile memory. In addition, while connected to E-UTRAN the ME and MME are capable of storing in volatile memory the NCC, NH and the related K_{ASME} used to compute keying material for the current EPS AS security context.
- Any successful run of an EPS AKA creates a partial native EPS security context. This context overwrites any existing non-current EPS security context.
- After a successful run of a NAS SMC relating to the eKSI associated with an EPS security context, this context becomes the current EPS security context and overwrites any existing current EPS security context.

The details of security context handling during inter-system idle mode mobility and handover are given in Section 4.17. During inter-system mobility from UTRAN to E-UTRAN, if ISR is not switched on for idle mode mobility, whenever a mapped context is created, it replaces the existing current EPS NAS security context in E-UTRAN. On creation of mapped context, the then current native EPS NAS security context becomes non-current and replaces any other existing non-current security context. A mapped EPS NAS security context can never be saved as non-current EPS NAS security context, i.e., in such occasion the mapped context is deleted. The current native EPS NAS security context may be removed when mobility happens from E-UTRAN to UTRAN.

4.9.2.3 Security context handling at TAU and attach request

Once the NAS security is activated by SMC (see Section 4.11), the NAS messages after that are integrity and optionally also confidentiality protected as activated during the NAS SMC. For TAU and Attach Request message integrity protection will be done using the current EPS NAS security context in the UE.

Below we present generalized level of security context handling on TAU and Attach Request; specific behavior is discussed in appropriate sections:

- TAU or Attach Request without any security protection:
  - The identity of the UE must be verified and AKA must be performed (see Section 4.10), else mapped context must be created (see Section 4.17).
- TAU or Attach Request with integrity protection:
  - The MME will accept the TAU or Attach Request if it can find the UE security context with the GUTI and KSI provided, i.e., the GUTI is from the given MME (see Section 4.10).
  - The MME passes the message, especially the TAU Request, as is to the old MME to which the GUTI points to and gets the security context from that old MME after successful integrity check (see Sections 4.10 and 4.12).
  - Failure in finding the context will lead to creation of new K_{ASME} by running AKA.
  - It is possible that the integrity protection is by mapped context for the TAU Request message; variations can happen in this case, details are given in Section 4.17 on inter-system mobility. Basic idea is to fetch the security context from SGSN and if the TAU Request message also indicates native EPS security context then it is brought in use.
- UE security capabilities is different in MME to that in the received TAU or Attach Request message.
  - Stored security context will be removed and new keys will be created using the received UE security capabilities; see Section 4.17.

4.9.3 NAS Counts and Handling

A NAS Uplink and Downlink Count is given to every NAS message sent by the UE (uplink) and MME (downlink). The NAS Count is always increased and never decreased; this provides replay protection and prevents recreation of exactly the same value for parameters dependent on NAS Count, e.g., K_{eNB}, CK′, IK′, CK_{SRVCC}, IK_{SRVCC}, and NAS Token.

NAS Count value is reset to zero (start value) only when a new K_{ASME} is created. The creation of new K_{ASME} only happens when (1) a successful AKA is run and a partial native EPS NAS security context is created and (2) on inter-system handover or idle mode mobility resulting in mapped EPS security context creation.

4.9.4 Cryptographic Algorithms and Usage Overview

Currently there are two confidentiality protection algorithms defined by 3GPP. These are the SNOW 3G and advanced encryption standard (AES). Another algorithm is the NULL algorithm that does not provide any security. Having two fundamentally different algorithms guarantees that at least one algorithm will be available for use if the other is cracked. A new algorithm known as ZUC is being defined by 3GPP due to regional requirements from China. All algorithms are based on 128 bits keys. TS 33.401 defines EEA0, EEA1 and EEA2 as the default confidentiality protection algorithms and, EIA1 and EIA2 as default integrity algorithms; explanation of the algorithms is given below.

4.9.4.1 Null algorithm

EPS security defines NULL algorithm for emergency calls for the cases where authentication and thus key generation is not possible; this is also known as the limited service mode (LSM). This NULL encryption and integrity algorithms are known as EPS encryption algorithm 0 (EEA0) and EPS integrity algorithm 0 (EIA0), respectively. The standard does not define how to implement these algorithms but does give the expected output.

The EEA0 is supposed to be implemented such that a keystream of zeroes is generated of the same length as the input parameter. Length is the only input parameter required for generation of the keystream. On the other hand the EIA0 generates MAC-I and XMAC-I of all zeroes with the length of 32 bits.

4.9.4.2 Input parameters for confidentiality and integrity

In this section, we discuss the input parameters for confidentiality and integrity protection. It is better to read this section together with the following two sections on confidentiality and integrity algorithms. See Section 4.2.2 for layers and their function for confidentiality and integrity protection.

Input parameters for confidentiality and integrity protection of a message are 128-bit integrity or confidentiality key (AS or NAS keys) as KEY, a 5-bit bearer identity BEARER, the direction of transmission DIRECTION, and a bearer specific, time- and direction-dependent 32-bit input COUNT. For AS the COUNT is equivalent to the 32-bit PDCP COUNT [TS 36.323] while for NAS the COUNT is constructed as follows:

$$\text{Count} := 0 \times 00 \,||\, \text{NAS OVERFLOW} \,||\, \text{NAS SQN}$$

where

- the leftmost 8 bits are padding bits including all zeros,
- NAS OVERFLOW is a 16-bit value which is incremented each time the NAS SQN is incremented from the maximum value, and
- NAS SQN is the 8-bit sequence number carried within each NAS message.

In case of confidentiality protection there is an additional input parameter, namely the length, LENGTH, of the key stream to be generated by the algorithm. For AS the BEARER corresponds to the radio bearer identity while for NAS it is equal to a constant value 0×00. The BEARER identity is not necessary for NAS since there is only one NAS signalling connection per pair of MME and UE, but is included as a constant value so that the input parameters for AS and NAS will be the same, which simplifies specification and implementation work.

4.9.4.3 Confidentiality algorithms

The two existing confidentiality protection algorithms, SNOW 3G and Advanced Encryption Standard (AES) in Counter (CTR) mode, are known as EEA1 and EEA2 respectively.

The input parameters to the confidentiality protection algorithm are a 128-bit key named as KEY, a 32-bit count, a 5-bit bearer identity BEARER, the 1-bit direction of the transmission, i.e., DIRECTION, and the length of the keystream required, i.e., LENGTH. The DIRECTION bit is 0 for uplink and 1 for downlink.

Figure 4.12 illustrates the use of the confidentiality protection algorithm EEA to encrypt plaintext by applying a keystream using a bit per bit binary addition of the plaintext and the keystream. The plaintext may be recovered by generating the same keystream using the same input parameters and applying a bit per bit binary addition with the ciphertext.

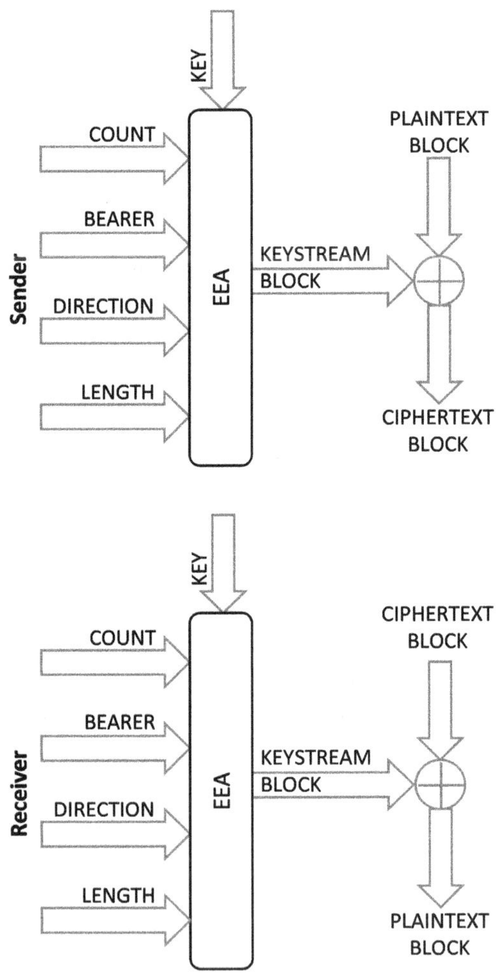

Fig. 4.12 Application of confidentiality protection algorithm.

Based on the input parameters, the algorithm generates the output keystream block KEYSTREAM, which is used to encrypt the input plaintext block PLAINTEXT to produce the output ciphertext block CIPHERTEXT.

The input parameter LENGTH will affect only the length of the KEYSTREAM BLOCK, not the actual bits in it.

4.9.4.4 Integrity algorithms

The two existing integrity algorithms, SNOW 3G and Advanced Encryption Standard (AES) in Cipher-based Message Authentication Code (CMAC) mode, are known as EIA1 and EIA2, respectively.

The input parameters to the integrity algorithm are a 128-bit integrity key named KEY, a 32-bit count, a 5-bit bearer identity called BEARER, the 1-bit direction of the transmission, i.e., DIRECTION, and the message itself, i.e., MESSAGE. The DIRECTION bit is 0 for uplink and 1 for downlink. The bit length of the MESSAGE is LENGTH.

Figure 4.13 illustrates the use of the integrity algorithm EIA to authenticate the integrity of messages.

Based on these input parameters, the sender computes a 32-bit message authentication code (MAC-I/NAS-MAC) using the integrity algorithm EIA. The message authentication code is then appended to the message when sent. For integrity protection algorithms other than EIA0 the receiver computes the expected message authentication code (XMAC-I/XNAS-MAC) on the message received in the same way as the sender computed its message authentication code on the message sent and verifies the data integrity of the message by comparing it to the received message authentication code, i.e., MAC-I/NAS-MAC.

The 16 least significant bits of the 32 bits NAS-MAC is appended to the NAS service request message.

4.9.4.5 Key derivation function

Throughout this chapter, we discuss about key derivation. All the key derivations are based on the key derivation function (KDF) explained in this section.

For EPS the KDF, the key derived is result of the function:

$$\text{derived key} = \text{HMAC-SHA-256 (Key, S)}$$

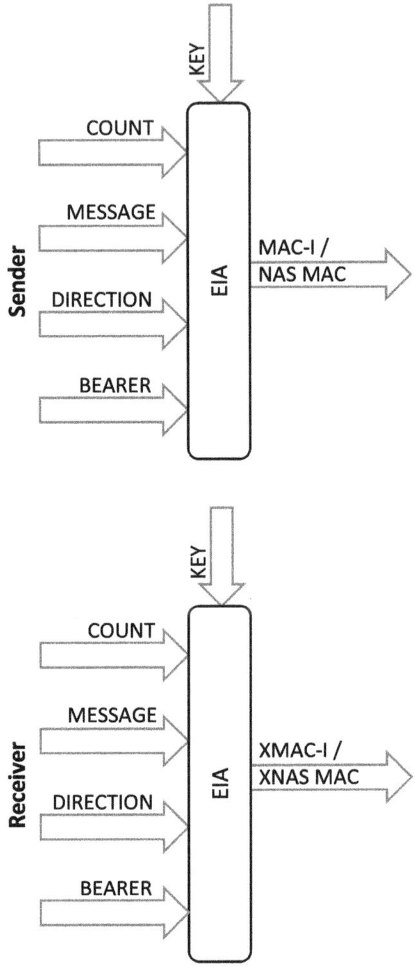

Fig. 4.13 Derivation of MAC-I/NAS-MAC (or XMAC-I/XNAS-MAC).

For each derivation, a key is defined together with the string S. The string S is a concatenation of one octet FC together with input parameters (P) and their length (L):

$$S = FC||P0||L0||P1||L1||P2||L2||P3||L3|| \cdots ||Pn||Ln$$

FC is used to distinguish between different instances of the algorithm. For example, for K_{ASME} derivation FC is 0×10 and for K_{eNB} derivation it is 0×11. In case of EPS the FC values is in the range of 0×10 to $0 \times 1F$.

4.10 Authentication and Key Agreement (AKA)

Authentication in EPS is the same as in the authentication part of the AKA in UMTS (see Figure 4.16), except that the key generation is different. In this section, we start looking at the cases when authentication should happen, after that we discuss the authentication procedure. In Section 4.11, we present the key generation and agreement between the network and the UE.

4.10.1 Basic Requirements

For EPS, it is necessary for the UE to allow usage of Rel. 99 or later version of USIM. Authentication is not possible if the UE has SIM. Thus a SIM-based UE cannot access EPS network. The AKA procedure for EPS is known as EPS AKA because the resulting key is K_{ASME} created from CK and IK as in UMTS AKA. A procedure known as security mode command (SMC) leads to key creation for AS and NAS from K_{ASME} (see Section 4.11).

4.10.2 Authentication Decision and Context Retrieval

In this section, we discuss other cases that lead to authentication, i.e., EPS AKA incase of EPS.

As a principle, the operator can decide when and how often authentication (AKA) should take place but a UE joining the network for the first time or without appropriate security context/association must also be authenticated (as in Figure 4.6). Besides these reasons, another case for AKA is when the NAS Count is about to wrap around; see Section 4.9.3.

The MME at first should identify the UE to find its security context. In case the UE cannot be identified or the associated security context is not available at the given MME then authentication must take place. In the ATTACH Request or TAU Request (see Sections 4.3 and 4.4), the GUTI (a temporary ID bound for the UE bound to a given MME) and KSI are sent, still different cases can happen:

- It is possible that the Attach Request or TAU Request are sent without any security protection and without any GUTI (and P-TMSI) and/or KSI. In such a case, the UE cannot be identified and thus authentication must take place. So as to authenticate without any prior UE identity information, the MME will request the UE for

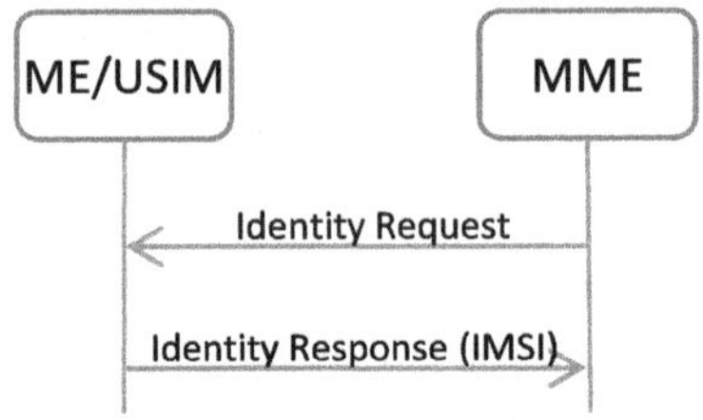

Fig. 4.14 User identity query.

its permanent identity in response to which the UE sends its IMSI (see Figure 4.14). The IMSI is sent in clear text, thus leading to breach in user identity confidentiality.

- The MME to which the UE connects to might not have the security context. In such case the MME will forward the TAU Request (or Attach Request) message to MME corresponding to the GUTI; this allows the old MME to check the integrity of the TAU Request, in response to which the old MME sends the security context to the new MME. Still in some cases the old MME might not have the security context. This will lead to MME requesting UE for its IMSI as discussed above, after which the authentication procedure can start.

The step of communication and context retrieval from old MME is also valid for MMEs in different serving network domains. The difference being that the unused EPS authentication vector (AV) or non-current EPS security context are not distributed between MMEs of different PLMNs.

It is possible that the given UE has or had moved from UMTS network. The rules in such case is that the UMTS context cannot be exchanged between MMEs. On the other hand, a SGSN may send unused AVs to the MME in the same serving network domain while the MME may also send UMTS AVs back to the same SGSN but not to any other SGSN.

4.10.3 Authentication Procedure

AKA is a secret key algorithm that provides mutual authentication between the USIM and the network, a long-term unique secret key K must be shared between the USIM and the authentication center (AuC). So as to distinguish EPS AKA from UMTS AKA a specific bit, known as separation bit, in the so called authentication management function (AMF) is set to 1 for EPS AKA,

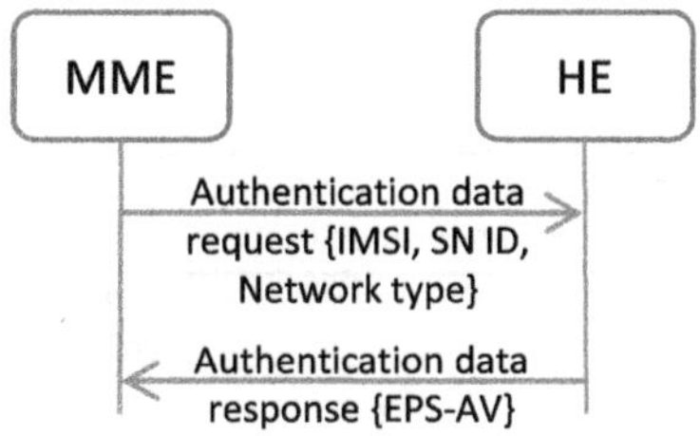

Fig. 4.15 Distribution of authentication vector from HE to MME.

this also means that the given bit cannot be used for operator specific purpose as was originally decided in 33.102 for UMTS.

It is the HSS that generates authentication vector and transfers them to the MME of the network under which coverage the subscriber's UE is located *t* the moment authentication is performed. The HSS generates the authentication vector on receiving the authentication data request from MME (see Figure 4.15).

The EPS authentication vector (AV) is based on the UMTS AV as given below except that the CK and IK are converted to K_{ASME} using serving network ID (SNID) as explained in Section 4.9. In case of EPS the key set identifier for the K_{ASME} (KSI_{ASME}) is also sent to the MME. The UMTS AV values are as follows:

- The challenge RAND
- The challenge response XRES
- The cipher key CK
- The integrity key IK
- The authentication token AUTN.

EPS AV values are as follows:

- The challenge RAND
- The challenge response XRES
- The key K_{ASME}
- The authentication token AUTN

The AKA procedure is outlined in the following steps, see associated Figure 4.16:

1. Authentication is initiated by the network when the UE is first switched on and due to reasons as explained in Section 4.10.2. The

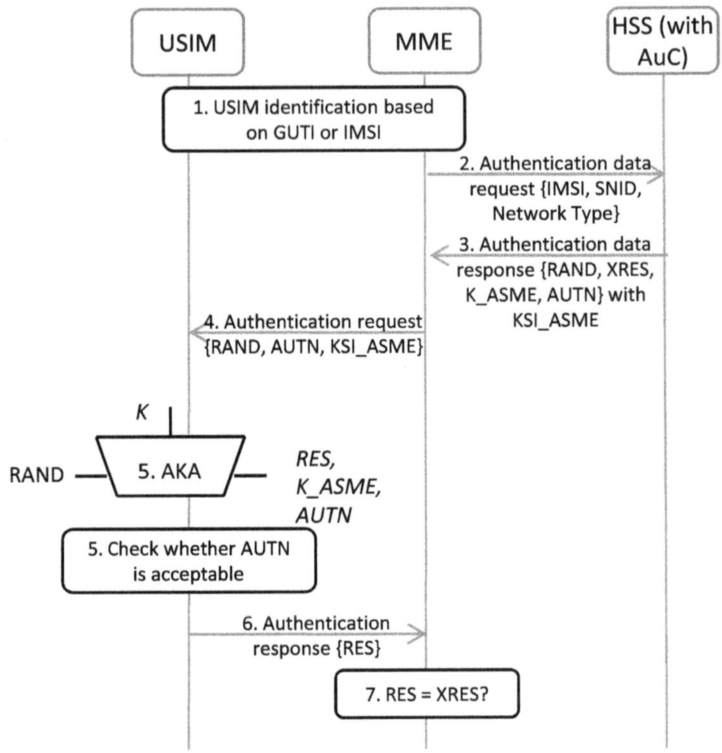

Fig. 4.16 EPS authentication procedure.

UE (USIM) sends its identity through GUTI or IMSI as discussed earlier in ATTACH Request or TAU Request message or else in the initial Service Request message.

2. The MME establishes the identity of the USIM through the GUTI. If the GUTI is recognized then the MME sends a request for authentication to the HSS, if not it will request the UE's IMSI as in Figure 4.14.
3. The AuC generates the UMTS AV by using the key secret K associated with the IMSI that identifies the USIM, then derives K_{ASME} and sends the AV with K_{ASME} (thus the EPS AV) and KSIASME (excluding CK and IK) to the MME.
4. The MME sends RAND and AUTN to the USIM.
5. The USIM runs AKA and generates AUTN then the USIM verifies if AUTN is acceptable, where AUTN is the network authentication

token. If AUTN is valid the USIM calculates the expected response XSRES using its secret key K and the challenge RAND. XSRES is sent to the MME for verification. The USIM also calculates the encryption key CK and the integrity key IK.
6. The USIM, via ME, then sends the Authentication response with RES to the MME.
7. If RES = XRES, then the USIM is authenticated and allowed access to the network. If RES ≠ XSRES then an authentication rejected signal is sent to the USIM and access to the network is denied.

The authentication algorithm was not standardized by the 3GPP organization because the architecture demands that every operator manages his users' authentication and sends the AV to the MME. Nevertheless, the reference algorithm MILENAGE was designed and is used by most operators for UMTS thus probably also for EPS.

4.11 Algorithm Negotiation and Security Activation

After authentication is completed it is possible to start the security by generating keys for NAS and AS. The procedure to generate the keys is known as the security mode command (SMC). During SMC the decision for the cryptographic algorithm to be used is also taken. Unlike UMTS, in EPS the SMC is run separately whenever there is a need to setup the NAS level and the AS level keys. In this section, the SMC procedure is discussed starting with the algorithm negotiation requirements.

4.11.1 Algorithm Negotiation Requirements

The algorithm negotiation requirements for EPS are given below:

(a) An active UE and a serving network will agree on algorithms for NAS and AS thus the keys generated will be:

  (i) AS keys:

  - For RRC confidentiality and RRC integrity protection (to be used between UE and eNB)

- For UP confidentiality protection (to be used between UE and eNB)

(ii) NAS confidentiality and integrity protection keys (to be used between UE and MME)

(b) The serving network selects the algorithms to use dependent on the UE security capabilities and the configured allowed list of security capabilities of the currently serving network entity.

(c) The same set of confidentiality and integrity algorithms will be supported by the UE both for AS and NAS.

(d) Each selected algorithm will be acknowledged to the UE in an integrity protected way such that the UE is ensured that the algorithm selection was not manipulated, i.e., that the UE security capabilities were not bidden down.

(e) The UE security capabilities the ME sent to the network will be repeated in an integrity protected NAS level message to the ME such that "bidding down attacks" against the UE's security capabilities can be detected by the ME. The UE security capabilities apply to both AS and NAS level security. Although this is a requirement set in 3GPP security activity, in general most of the capabilities of UE are standard and therefore it does not need to be sent. Only UE capabilities that could differ per UE need to be sent to the network. At the time of writing this book, there were two mandatory cryptographic algorithms thus security-related UE capability was not required to be sent. A third algorithm, as discussed, earlier will be standardized by the end of 2010 or in 2011 that might be optional thus necessitating the transfer of cryptographic algorithm.

(f) Separate AS and NAS level security mode command procedures are required. AS level security mode command procedure configures AS security (RRC and UP) and NAS level security mode command procedure configures NAS security.

(g) Both integrity protection and confidentiality for RRC are activated within the same AS SMC procedure, but not necessarily within the same message.

(h) User plane confidentiality protection is activated at the same time as that of RRC.

(i) It will be possible that the selected AS and NAS algorithms are different at a given point of time.

4.11.2 Non-access Stratum

Each MME is setup with the non-access stratum (NAS) confidentiality and integrity protection algorithms in the order of priority chosen by the operator. Thus the MME chooses the algorithm with highest priority that is also in UE capability list and starts the NAS SMC to setup the NAS security context. At every S1 handover, that usually means change in MME, the NAS algorithms can be changed.

A successful NAS SMC message sequence (a round-trip) is given in Figure 4.17.

1. The MME first chooses the NAS algorithms, creates keys, and starts integrity protection.
2. The MME now sends the integrity protected (NAS-MAC) NAS security mode command to the ME with the eKSI, identifying

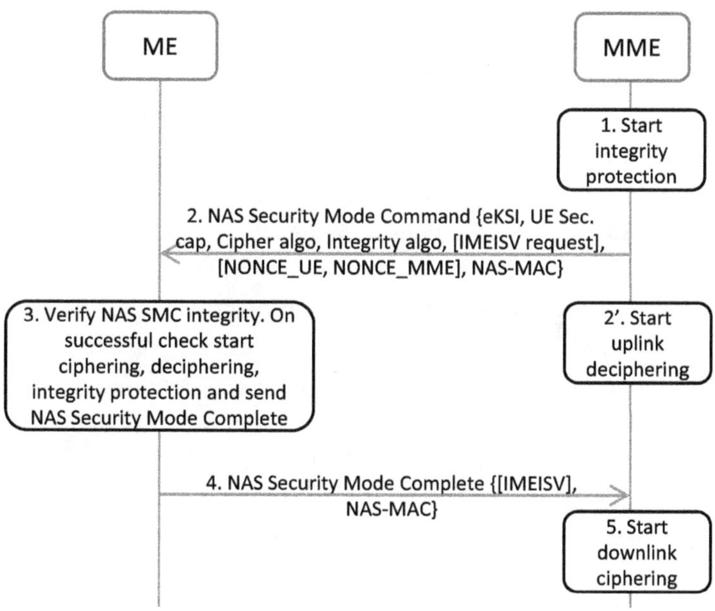

Fig. 4.17 NAS security mode command procedure.

the K_{ASME} used to derive the integrity protection key, replayed UE security capabilities, the selected NAS algorithms, and, $NONCE_{UE}$ and and $NONCE_{MME}$. The NONCEs are used to create mapped EPS context in inter-system (UMTS to LTE) idle mode mobility.

The MME receives the $NONCE_{UE}$ in the TAU Request message sent at the idle mode mobility to EPS network. Inter-system mobility aspects are discussed in Section 4.17.

After sending the NAS security mode command, the MME starts deciphering of uplink (message from the UE to the network) messages (see 2′ in the figure).

3. The ME performs integrity verification of the received security mode command from the MME and on successful check the ME starts ciphering, deciphering and integrity protection.

   UE capabilities and $NONCE_{UE}$ are sent to the ME to verify whether the MME had received the correct information previously, which could be sent in message without integrity protected. Integrity check by the ME of the NAS security mode command from MME proves that the message was not modified by anyone.
4. The ME sends NAS security mode complete message to the MME that is protected by the chosen algorithms in the message from MME (message 2). The protection of the response from ME will depend on the decision taken by the MME, i.e., mandatory integrity protection (NAS-MAC) and optional confidentiality protection with the algorithm chosen by the MME. The choice of no-confidentiality protection can be taken by the MME by sending the NULL algorithm as the chosen confidentiality protection algorithm. Confidentiality and integrity protection starts before sending the NAS security mode complete, particularly ciphering starts at this given point because the complete message might contain the IMEISV, see Section 4.7, that should not be sent in clear.
5. Now the MME starts downlink ciphering.

All NAS messages from here on will be secured by the algorithms chosen during the NAS SMC except for messages those can be received or sent without integrity or confidentiality protection [TS 24.301]. In case the integrity check of the message from MME fails then the UE will send a NAS security mode reject message to the MME protected with the old keys and algorithm; as old context might not exist in the MME, the reject message might not be protected.

4.11.3 Access Stratum

In the case of access stratum (AS), the keys can be modified during handover without running a AS SMC. Thus the situation is different than NAS key handling, the reason being that there is far more handover between eNBs than that between different MMEs. Thus the overall impact will be much higher if SMC round-trip is needed at every inter-eNB handover.

As an initial step, every eNB is set with priority list of algorithms just like the MME. Initially the MME informs UE capabilities to the eNB; the eNB uses this to determine the cryptographic algorithms to be used with the given UE. In case of X2 handover the Source eNB informs the Target eNB about the algorithm while for S1 handover the information comes from the MME. The Target eNB informs the chosen algorithm in the handover command to the UE; this information of chosen algorithm is sent by Target eNB over the Source eNB to the UE secured by the algorithms applied by the Source eNB. The Target eNB also informs the MME about the UE security capabilities it received from the Source eNB so that the MME can take action if there is difference in UE capabilities that it has. Algorithm change is not required when the handover type is intra-eNB.

AS SMC takes place on indication from the MME. The AS SMC round trip is depicted in Figure 4.18. One of the important differences to NAS SMC is that the confidentiality protection at UE starts only after successful AS SMC. The reject message from ME is neither confidentiality nor integrity protected. The steps shown in Figure 4.18 are explained below:

1. The eNB starts integrity protection by choosing a common algorithm with UE that has the highest priority.
2. The eNB now sends the integrity protected (MAC-I) AS Security Mode Command to the ME together with the selected AS algorithms. After sending the AS Security Mode Command, the eNB starts downlink ciphering of UP and RRC messages (see 2′ in the figure).
3. The ME performs integrity verification of the received AS Security Mode Command from the eNB and on successful check the

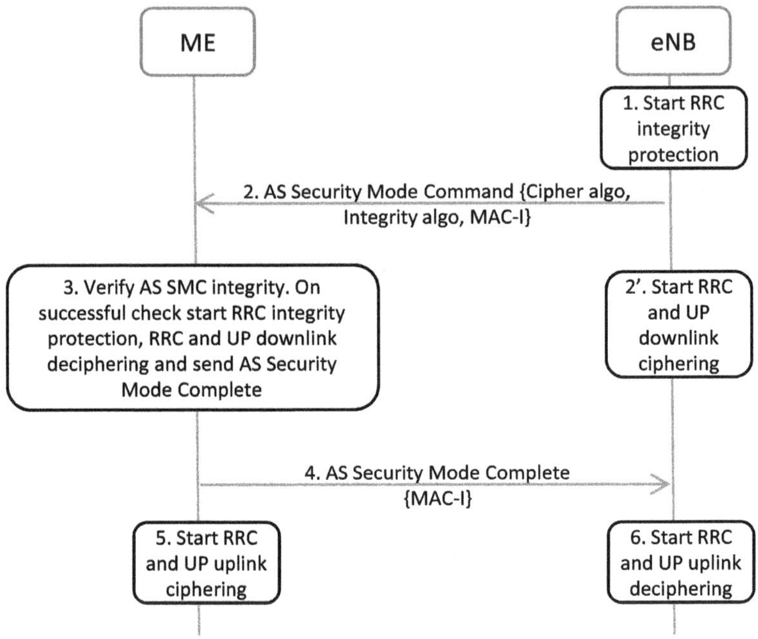

Fig. 4.18 AS Security mode command procedure.

ME starts RRC integrity protection and, RRC and UP downlink deciphering.

4. The ME then sends a AS security mode complete message to the eNB that is integrity protected by the algorithm chosen by the eNB (MAC-I).
5. Now the eNB starts RRC and UP uplink ciphering.
6. The eNB then starts RRC and UP uplink deciphering.

4.12 Key Handling at State Change

In this section, we cover the key handling on transition to or from the 4 states of NAS layer discussed in Section 4.2.2. Starting with discussion on transition to EMM-DEREGISTERED, then transition from EMM-DEREGISTERED to EMM-REGISTERED and ECM-CONNECTED, going on to transition to ECM-CONNECTED from ECM-IDLE and finally to ECM-IDLE to ECM-CONNECTED.

There are several different reasons that can lead to the transition to the EMM-DEREGISTERED state. The basic idea is that even if the UE or MME get into EMM-DEREGISTERED state when going back to EMM-REGISTERED state there should be means to avoid AKA. So as to avoid AKA the solution is to store the most recent full native EPS NAS security context in both the MME and UE; all other EPS NAS context are deleted. The most current native EPS NAS context will be in non-current state only if there is a current mapped EPS NAS context in such case the mapped context is removed and the native context is transitioned to and stored in the current state. In the case of UE the context should be stored in the USIM if it is capable to store EPS context else it is stored in the non-volatile memory (NVM) of the ME. So as to avoid data over the USIM and ME interface the K_{NASint} and K_{NASenc} are not stored (also in NVM), the reason being that these two keys can be recreated if the used algorithm and K_{ASME} are known. For MME the remaining AVs, if any, are also stored together with the UE capabilities. Specific causes like attach reject and HSS-initiated "subscription withdrawn" mean that all context in UE and MME are removed. There can be several reasons for state change to EMM-DEREGISTERED but in general the behavior given above will be followed.

The change in state from EMM-DEREGISTERED to EMM-REGISTERED or ECM-CONNECTED happens with a NAS attach request from the UE to the MME thus full (NAS and AS) EPS security context should be created if not existing. As the NAS keys are not stored in UE, before moving to ECM-CONNECTED state the UE must create the K_{NASint} and K_{NASenc}. On receiving the attach request message, the MME verifies whether it has the context used to integrity protect the message, certainly this does not happen if the message is not integrity protected. The MME checks for the context used, as given in Section 4.10.2, and decides to start AKA. After AKA the MME runs the NAS SMC procedure to get the new keys in use followed by AS SMC. When the MME has the context but it is in non-current state then the MME changes the state of the given context to current. As the AS keys are removed in EMM-DERIGESTERED state, the eNB starts an AS SMC to activate the AS context by using the most recent NAS uplink context.

On ECM-CONNECTED to ECM-IDLE transitions, the eNB no longer needs to store state information about the UE; thus all radio bearers of the UE are deleted leading to removal of AS security context both in the UE and eNB.

The MME and UE keep the newest full EPS NAS context. Here we use the term newest full EPS NAS context because it is possible that a full EPS NAS security context is created but the newly created K_{ASME} is not yet taken in use for AS keys thereby requiring the existence of the old K_{ASME} in parallel to the new one for AS keys when in ECM-CONNECTED state.

ECM-IDLE to ECM-CONNECTED state transition happens with the UE sending a message to the MME. The first thing at the MME is to check the existence of the security context and check the necessity for starting AKA, as discussed earlier in this section. The MME might also decide to perform AKA due to operator policy, e.g., an earlier handover from UMTS to LTE. The AS keys are also created if there is need of it, which is determined when MME receives NAS service request message or TAU request message with the active flag set by the UE.

Whenever the state transition to ECM-CONNECTED happens and the NAS SMC is not needed, the most recent NAS Uplink Count is used to create the K_{eNB} that is sent to the eNB by MME. The MME further initializes the value of the next hop chaining counter (NCC) to zero and derives a next hop (NH) parameter as specified using the newly derived K_{eNB} and the K_{ASME} as the basis for the derivation. This fresh $\{NH, NCC = 1\}$ pair is stored in the MME and used for the next forward security key derivation. The details of NH, NCC, and forward security is discussed in Section 4.13. The K_{eNB} is then used to derive the AS keys in both eNB and UE with the run of AS SMC.

As stated in Section 4.4, when TAU is sent the UE state is changed to ECM-CONNECTED. The UE uses the current EPS security context to integrity protect the TAU Request that includes the corresponding GUTI and eKSI value. The TAU Request is not confidentiality-protected. If at MME the eKSI and GUTI indicates that non-current EPS NAS context is used then the MME changes the state of the given context to current. Further, the AS context should only created if the "active flag" is set in the TAU request, thereby indicating that the UE is active and radio bearers need to be established although it could be different in practice. To derive the K_{eNB} without a NAS SMC run, the NAS Uplink Count of the TAU request message sent from the UE to the MME is used. In the case an AKA is run successfully, the uplink and downlink NAS Count will be set to the start values (i.e., zero). If there is a NAS security mode command after the TAU request but before the AS SMC, the UE and MME use the NAS Count of the most recent NAS security mode complete (i.e., the NAS

Uplink Count) and the related K_{ASME} as the parameter in the derivation of the K_{eNB}. From this K_{eNB} the RRC protection keys and the UP protection keys are derived. The reason of NAS SMC run might be due to different algorithm use in the MME where the UE is currently attached to.

4.13 Handover Key Handling

As discussed earlier in the chapter, the confidentiality and integrity protection in EPS ends at the eNB. This brings strong security requirements on the eNB security itself but even if such security enhancements are done, eNB is the weakest link as it will be installed in locations accessible to intruders. In case an attacker compromises a eNB then the threat from such attack should be contained, i.e., the attacker might be in control of the keys of UEs at the given eNB but having such knowledge must not lead to compromise of future communication of the UE. Therefore it is important to have a forward security solution for handover, i.e., knowledge of UE key at any given time should not lead to the attacker being able to decrypt any user communication in future. Thus a key handling solution providing forward security for AS was developed, whose general model is shown in Figure 4.19 and the basic message sequence is shown in Figure 4.20. Detailed explanation of handover and key handling is given below.

To achieve forward security it is necessary to (1) generate new key for the UE at handover, (2) the key should be derived using some random value, and (3) random value used to generate the key should not be sent over the air because it might be sent without security or by using the keys at Source eNB that is assumed to be compromised. The random value can be generated at the eNB or MME in the network side but normally eNBs can be compromised therefore the obvious choice is to generate the random value in the MME. Also, the random value can be generated (1) before handover, (2) during handover, or (3) after handover. Generation of random value after handover means that at least the security for the first handover (first hop) can be compromised if the Source eNB during the first handover was compromised. Still this is the solution accepted by 3GPP because it is the best possible compromise to security and complexity. In EPS, the forward security at handover is achieved by using the so called Next-Hop (NH) parameter. The initial NH is derived using the initial K_{eNB} derived after AKA; subsequent NH is derived using previous NH with the

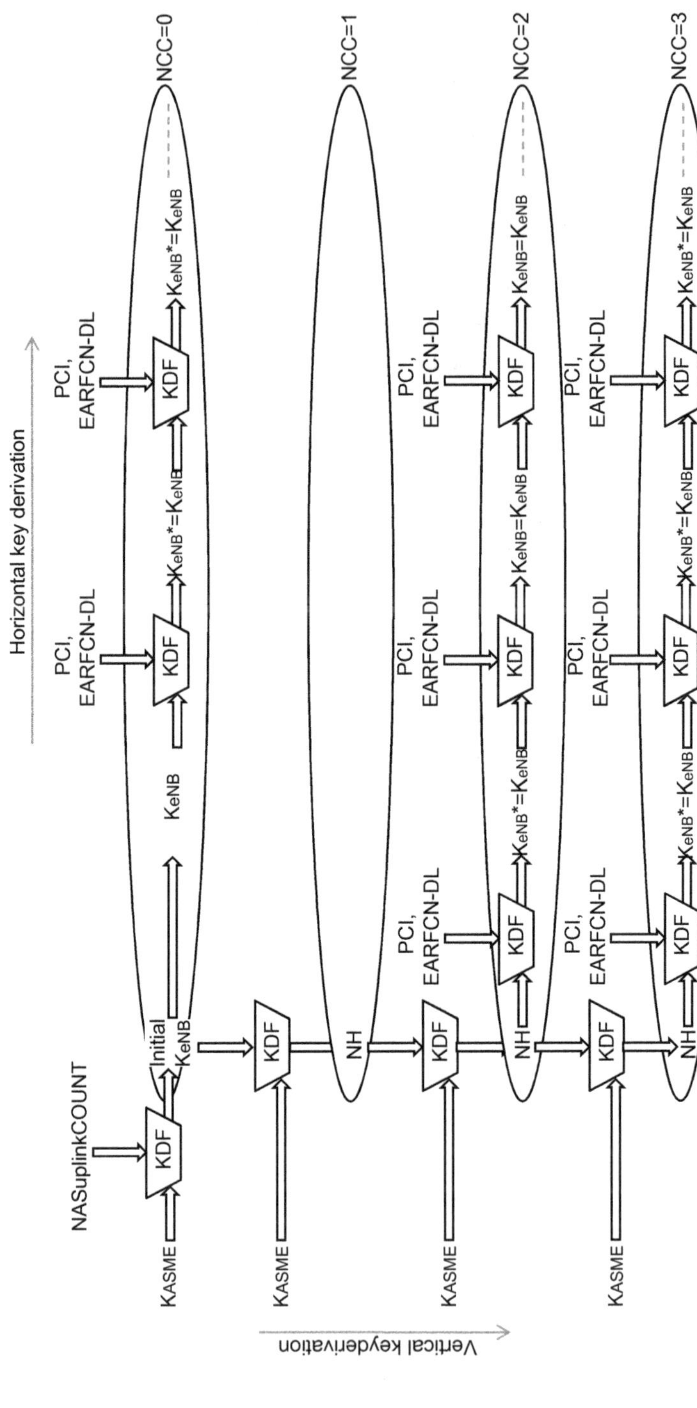

Fig. 4.19 Model for the handover key chaining.

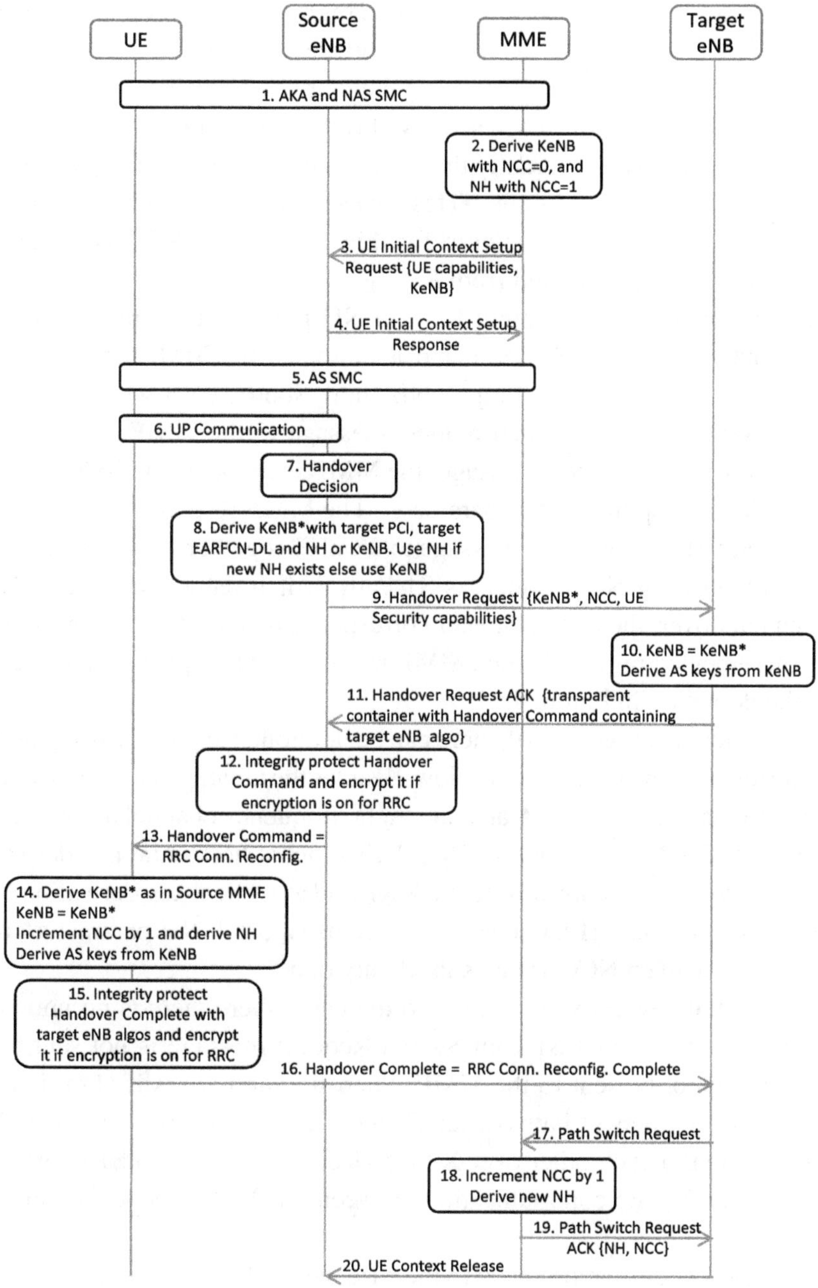

Fig. 4.20 Basic handover message sequence with security aspects.

256-bit K_{ASME} as the input key in the KDF. This results in a NH chain, where the next NH is always fresh and derived from the previous NH.

A NH corresponds to NH chaining counter (NCC) value; the NCC value is associated to the K_{eNB} that was derived from NH parameter corresponding to the given NCC value. Initially the K_{eNB} is derived from the K_{ASME} and the most recent NAS Uplink Count; NH is not used for this initial K_{eNB} derivation and NCC is considered to be zero for this virtual NH. NCC is set to zero whenever the K_{eNB} is derived from K_{ASME}.

At the creation of the initial K_{eNB}, a NH parameter is created from the K_{eNB} corresponding to NCC = 1 and is stored in the MME and UE. After a handover, the previously Target eNB (now Source eNB where the UE is located) sends S1 Path Switch request message to the MME. On receiving the S1 Path Switch request message, the MME increments the NCC by 1 and derives the corresponding NH parameter. The MME responds with a S1 Path Switch request acknowledge message to the eNB together with NH parameter and corresponding NCC value. As S1 Path Switch request message is only sent on handover, the NH parameter corresponding to NCC = 1 is never used (it is incremented by 1 = 2 when MME receives the first path switch request) thus the depiction in Figure 4.19.

As depicted in Figure 4.19, there can be horizontal and vertical key derivation during handover; explained below. Besides the target physical cell identity (PCI) and the target E-UTRA absolute radio frequency channel number-down link (EARFCN-DL) as input to K^*_{eNB} derivation, at horizontal key derivation K^*_{eNB} is derived from currently active K_{eNB} and at vertical key derivation K^*_{eNB} is derived from the NH parameter in place of K_{eNB}. A NH parameter corresponding to a given NCC value is used only once.

Horizontal key derivation refers to the case where there is no unused or new NH parameter and S1 Path Switch Request message is not sent to the MME or cannot be sent to the MME. Such situation arises if (1) it is intra-eNB handover when S1 Path Switch Request message is not sent or (2) eNBs are connected to each other over X2 interface but there is no S1 interface to the MME or (3) the communication between the MME and eNB somehow fails.

Vertical key derivation refers to the case where the MME was able to send the new NH parameter and NCC value to the eNB. This key derivation happens whenever there is a new or unused NH parameter and connectivity with MME exists for X2 handover and of course also for S1 handover. The NCC value used is sent to the UE in the Handover Command message.

It is possible for the Source eNB to create K^*_{eNB} for several neighboring eNBs and sends this K^*_{eNB} with corresponding NCC to them. Such method helps when handover to a given eNB fails and the UE tries to connect to other neighboring eNB.

In the following, we present the unique points regarding three handover types, intra-eNB, X2 and S1 handovers (see Figure 4.18, 4.19 and 4.20).

- Intra-eNB: The derivation of K^*_{eNB} happens as discussed above. After handover the K^*_{eNB} is used as K_{eNB}. The eNB sends the NCC used for K_{eNB} to the UE in handover command message.
- X2 Handover: The derivation of K^*_{eNB} happens as discussed above. The Source eNB forwards the {K^*_{eNB}, NCC} pair to the Target eNB which takes the KeNB* as KeNB and associates the NCC to it. The Target eNB then sends the NCC value to the Source eNB in the handover command message, which forwards the value to UE. On completion of handover signaling with the UE, the Target eNB sends the S1 Path Switch request message to the MME where the NCC is incremented by 1 and a new NH value is created, and the {NH, NCC} pair is sent to the eNB. The eNB replaces any other {NH, NCC} pair with the received pair. As the Source eNB creates the NH value, an attacker at a compromised eNB will know the key of a given UE at the Target eNB thus forward security is achieved only after handover away from the Target eNB, i.e., 2 hop.
- S1 Handover: In the case of S1-Handover, the Handover Required message from the Source eNB triggers the MME to increment the NCC by one and derive a new NH parameter. The source MME stores the newly derived {NH, NCC} pair and sends it to the target MME (the source and target MME can be the same) in the S10 Forward Relocation Request message together with the K_{ASME} used to compute the {NH, NCC} pair and the corresponding eKSI. The target MME stores the {NH, NCC} pair and sends it to the Target eNB within the S1 Handover Request message. The Target eNB in turn derives the K_{eNB} with the new {NH, NCC} pair together with PCI and its frequency EARFCN-DL. The Target eNB associates

the NCC value received from MME with the K_{eNB}, sends the NCC value to UE in the handover command sent via the Source eNB, and removes any old {NH, NCC} pair. For S1-handover, the Source eNB will include AS algorithms used in the source cell (confidentiality and integrity algorithms) in the message to the Target eNB. The AS algorithms used in the Source eNB are provided to the Target eNB so that the Target eNB can decipher and integrity verify the RRCReestablishmentComplete message from the UE in potential RRCConnectionReestablishment procedure. In case of NAS algorithm change at S1 handover, i.e., the new MME has other cryptographic algorithm as higher priority, NAS SMC should take place.

Explanation of the messages in Figure 4.20 is given below based on discussion above:

1. The initial step is to perform AKA and NAS SMC; see Sections 4.10 and 4.11 respectively
2. The MME, that is serving the UE, now derives the K_{eNB} with $NCC = 0$. Here NCC is 0 because this is the initial K_{eNB} and now NH is associated to it. The MME also derives a NH associated with the $NCC = 1$.
3. The MME then sends a UE Context Setup Request message to the Source eNB together with the UE capabilities and the K_{eNB}.
4. The Source eNB acknowledges with a UE Initial Context Setup Response to the MME.
5. The usual AS SMC now takes place; see Section 4.11, whereby AS keys are created using the initial K_{eNB}.
6. Any UP communication can now take place.
7. Due to measurement reports, the Source eNB can now take a handover decision.
8. On taking the handover decision, the Source eNB derives the K^*_{eNB}.
9. The Source eNB then sends the K^*_{eNB} to the Target eNB.
10. The Target eNB takes the K^*_{eNB} as K_{eNB} and derives the AS keys.
11. Then the Target eNB sends a Handover Request ACK message to the Source eNB that contains a handover command message

(RRC connection reconfiguration) that in turn contains the algorithms to be used at the Target eNB. The handover command message is carried in a so called transparent container, which means that the Source eNB will not perform any actions with the content of the message and will forward the container to the UE.

12. The Source eNB then performs integrity protection of the transparent container and encryption if RRC confidentiality protection was switched on at AS SMC.
13. This secured handover command is now sent by the Source eNB to the UE.
14. On receiving the Handover Command, the UE derives K^{*}_{eNB} and takes it as K_{eNB} from which the UE also derives the AS keys. The UE also increments NCC by 1 and derives the new NH value.
15. Now the UE will start communicating with the Target eNB. The UE therefore integrity protects, and optionally confidentially protects, the handover complete message.
16. The UE then sends the handover complete (RRC connection reconfigure complete) message to the Target eNB.
17. On receiving handover completion message from the UE, the Target eNB sends a Path Switch message to the MME to inform the new location of the UE.
18. The MME in turn, like UE, increments the NCC by 1 and derives a new NH value.
19. Now the MME sends a Path Switch request ACK (Acknowledgement) message to Target eNB together with the new NH and NCC value that will be used for future handover.
20. As completion of handover the Target eNB now sends the UE context release message to Source eNB.

4.14 Key Change On-The-Fly

Key change on-the-fly was developed in 3GPP to allow change in key when different security context needs to be activated. For AS the activation would happen without using the SMC.

For complete key hierarchy re-keying, one has to generate a new K_{ASME} and then all other keys lower in the key hierarchy. For NAS key change, a

NAS SMC is done after AKA or other non-current native EPS AKA context is taken in use. The reason for taking non-current EPS AKA context in use is that potential NAS Downlink or Uplink Count of the current security context is about to wrap around, which normally would be the mapped context because a full native EPS context will not be available in both current and non-current states. This also means that the non-current EPS AKA context should have sufficiently low NAS Downlink and Uplink Count values.

In case of AS keys, the rekeying is done based on intracell handover procedure explained in Section 4.13. The reason for key change could be: (1) the PDCP count wrap around (PDCP maintains a 32-bit count value for a given radio bearer identity and K_{eNB}), (2) due to AKA run and activation of the new K_{ASME}, (3) activation of native context after handover from UTRAN or GERAN. For the case of PDCP count wrap-around, a intracell handover procedure is sufficient but for rest of the two cases there are some differences as given below. As for NH, the old K_{ASME} is used until context modification is completed after which the new K_{ASME} is used.

For the case of AKA run, the new K_{eNB} is sent to the eNB in the UE CONTEXT MODIFICATION REQUEST message triggering the eNB to perform the rekeying. The eNB in this case indicates UE that a key change on-the-fly is taking place, the procedure used here is the same as intracell handover. There could also be indication that the eKSI has changed, in this case the UE derives a temporary K_{eNB} using the uplink NAS Count in the most recent NAS Security Mode Complete message and the new K_{ASME} as input. From this temporary K_{eNB} the UE will derive the K^*_{eNB}. The eNB will take the K_{eNB} it received from the MME, which is equal to the temporary K_{eNB}, as basis for its K^*_{eNB} derivations. From this step onwards, the key derivations continue as in a normal handover. If the AS level rekeying fails, then the MME will complete another NAS security mode procedure before initiating a new AS level rekeying. This ensures that a fresh K_{eNB} is used.

For the case of rekeying initiated by the MME to reactivate a non-current full native EPS security context after handover from GERAN or UTRAN the same procedure as above applies.

4.15 Connection Reestablishment: Token

At times, a handover can fail or something similar can happen, leading to the UE attempting to connect to another cell of a given eNB. Thus at least cells in Target eNB and potentially those in Source eNB should be prepared for the UE when it tries to reconnect with the network after failed handover by using RRCConnectionReestablishment procedure. The RRCConnectionReestablishment procedure is over SRB0, a bearer that is not integrity-protected and thus prone to attack as the receiving side can neither authenticate the message nor check whether it was modified. Therefore an authentication mechanism is needed when the UE tries to reconnect to a different eNB, or different cell of the same eNB.

The basic idea is that the Source eNB prepares a token and K^{*}_{eNB} for all cells in Target eNB and optionally also for its cells. The Source eNB then sends security context containing K^{*}_{eNB}s and tokens for each cell to be prepared, as well as the corresponding NCC, the UE EPS security capabilities, and the security algorithms used in the source cell for computing the token, to the Target eNB.

The token is 16 least significant bits of the output of integrity protection, see Figure 4.13, using the negotiated EIA of the "Message" source C-RNTI, source PCI and target Cell-ID. The other parameters are as follows:

- KEY will be set to K_{RRCint} of the source cell
- All BEARER bits will be set to 1
- DIRECTION bit will be set to 1
- All Count bits will be set to 1.

The K_{eNB} of the source cell is kept until handover is completed successfully be it with a reconnection attempt. For S1-handover the K^{*}_{eNB} is removed by the receiving eNB. The UE sends the RRCConnectionReestablishmentRequest message to one of the target cells of the Target eNB together with the token. The Target eNB checks the validity of the token and responds with a RRCConnectionRestablishment message with the NCC or else with a RRCConnectionRestablishmentReject message. The UE uses the NCC value to derive the associated NH parameter and then it derives the K^{*}_{eNB}, which is used as K_{eNB}, with input parameters consisting of NH value, physical cell ID (PCI) and the EARFCN-DL as discussed earlier. The RRCConnectionReestablishmentComplete message is sent over the radio bearer called SRB1, which is integrity and confidentiality protected using the AS algorithms selected for the Source eNB and the new keys. The RRCConnectionReconfiguration procedure used to reestablish the remaining radio bearers only include integrity protected and ciphered messages. AS algorithms for the Target eNB is negotiated and used after the completion of RRCConnectionReestablishment procedure.

4.16 Local Authentication

An optional local authentication procedure for eNB is defined in 3GPP similar to that in UMTS. The trigger for the local authentication is the threshold set for PDCP count associated to each radio bearer. All messages in this procedure are integrity protected and the procedure is shown in Figure 4.21. The procedure starts with the eNB sending a counter-check message to the UE with most significant parts of the PDCP count values reflecting the amount of data sent and received for each active radio bearer. The UE compares the values received from the eNB and responds back with a counter-check response message that carries any UE PDCP Count value that was different than that sent by the eNB. If there is no value in the response message then the eNB does not do anything but takes action set by the operator if it receives a message with different PDCP Count. The action can be informing MME or O&M server and dropping the given bearer.

4.17 Inter-System Mobility

EPS system will not provide full coverage from day one thus there is a necessity to provide mobility with deployed 3GPP systems, like UMTS and GERAN. In this section, we present the solution in 3GPP TS 33.401 for such mobility. The mobility is either idle mode mobility or handover and depends on the

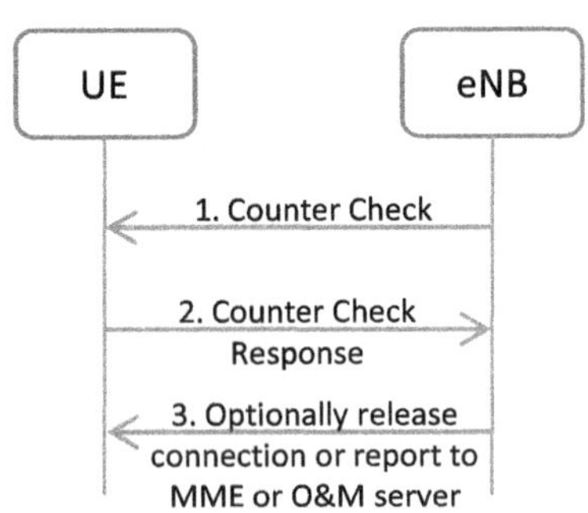

Fig. 4.21 eNB periodic local authentication procedure.

availability of security context of the target system at the UE. We first present the inter-system idle mode mobility and then move over to handover.

4.17.1 Idle-Mode Mobility

4.17.1.1 E-UTRAN to UTRAN

We start the discussion of idle-mode mobility from E-UTRAN to UTRAN; see Figure 4.22 for basic message sequence. In UMTS there is a routing area update (RAU) procedure similar to the TAU in EPS (see Section 4.4). Obviously the RAU from UE should identify the UE and its context in the network. So the UE, if it has been in UMTS previously and has the UMTS

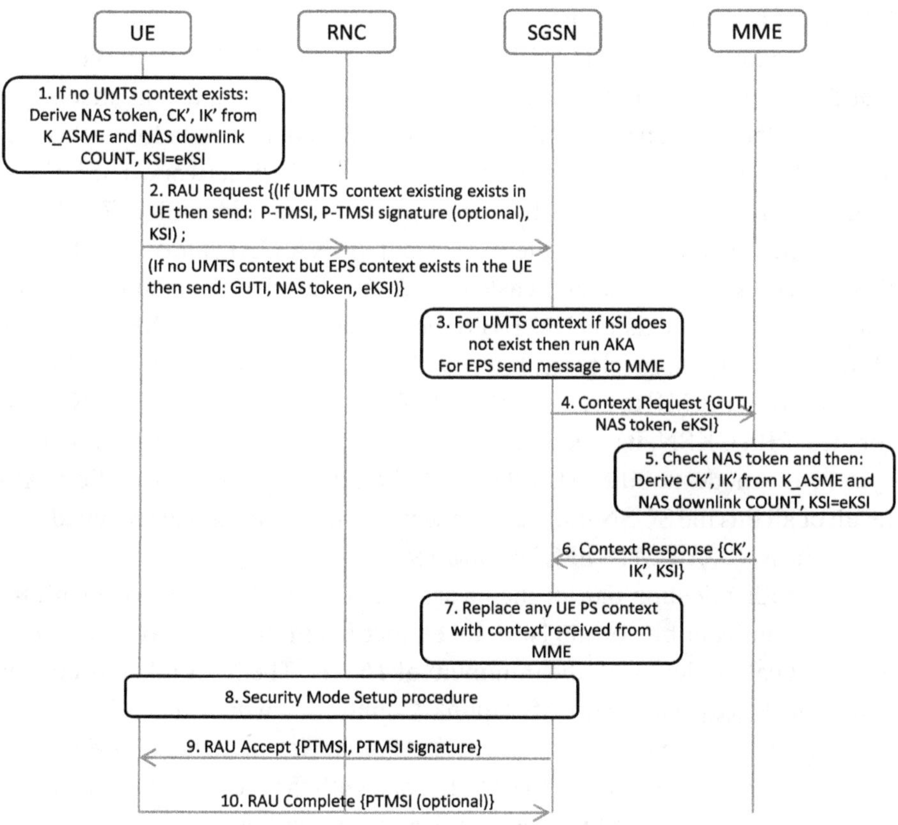

Fig. 4.22 Basic message sequence for E-UTRAN to UTRAN idle mode mobility.

security context stored in its USIM, will inform its identity to the network by sending the P-TMSI (see Section 4.7) and its security context by sending the KSI. In UMTS there is also an optional concept of P-TMSI signature of 3 Octet that is assigned by SGSN and used by SGSN to authenticate a P-TMSI or the UE to authenticate the network assigning it. Having P-TMSI signature prevents running of AKA everytime the UE sends a RAU. If a UE is assigned a P-TMSI signature then it must send the signature in the RAU together with the P-TMSI. In case a valid security context for the given UE cannot be found by the network then an AKA will be run. Note that the P-TMSI is sent in the old-P-TMSI information element (IE) of the RAU.

It is also possible that the UE does not have the UMTS security context, in such case the solution is to create security context from the EPS security context; this is known as the mapped security context. Once again the UE should inform the network about the security context available with it; this is done by the UE sending GUTI in the old-P-TMSI IE of the RAU. Then the issue comes that the given UE should be authorized by the network to prevent attack by a rogue UE; the solution is to send a NAS token (see explanation below in this section) in the RAU that is passed on to the old-MME in EPS network. The UE also sends the eKSI in the RAU so that the security context is identified. The MME derives the CK′ and IK′ from the K_{ASME} and the NAS Downlink Count value; this mapped key takes the eKSI as the KSI. The thus derived mapped security context replaces all existing UMTS PS security context in the SGSN and the UE (ME and USIM); this avoids any future authentication. The thus created CK′ and IK′ is used by the ME to create the GPRS Kc, which is assigned the eKSI value for the GPRS CKSN. It is recommended that in the case of inter-operator inter-system idle mode mobility a UMTS AKA should take place at the earliest. As a result of all this the SGSN transfers the security algorithm and the mapped context to the RNC. A UMTS SMC is run after this.

The authentication part of the message to the old-MME is done with the x-least significant bits of the NAS-token that fits in the x bits available in P-TMSI signature field with the minimum of 16 bits. The NAS token is derived using the K_{ASME} and the NAS Uplink Count after which the NAS Uplink Count is increased by 1 in the UE. This NAS token is sent to the old-MME that creates the same and matches the created one with the received. On successful verification, i.e., thus leading to authentication and authorization of the identity of the UE sent, the old-MME derives and sends the CK′ and IK′ together with eKSI (and other needed information) to the SGSN. It is possible that the MME does not have the latest NAS Uplink Count, in such case the MME will check the received NAS token against tokens generated with the current NAS Uplink Count upto current NAS Uplink Count + L where L is set by the operator. The MME stores the NAS Uplink Count that leads to successful verification of the received NAS token. Note that the MME does not accept the same NAS token twice except for the case of retransmission during same mobility.

Based on the above discussion the message sequence in Figure 4.22 is explained below:

1. Before sending a RAU the UE checks whether it has the UMTS context and if it does not have the UMTS context then the UE derives the mapped context, CK′ and IK′, from K_{ASME} and NAS Downlink Count. The UE also sets the KSI in UMTS to eKSI.
2. Next the UE sends the RAU Request to SGSN via RNC with the UMTS context identification information — P-TMSI and KSI — if it exists; else the UE sends the EPS context identification information — GUTI with eKSI. The UE also sends information to authorize itself in the form of P-TMSI Signature when it comes to UMTS and NAS Token when it comes to EPS context.
3. The SGSN checks the RAU Request, if UMTS security context for the related KSI is missing then the SGSN runs UMTS AKA. If all information UMTS security context exists in the SGSN then the system can go on to step 8.

   In case the RAU Request has EPS context identification then the SGSN sends request for context to the MME.
4. The SGSN sends Context Request message to MME with GUTI, NAS Token and eKSI.
5. Now the MME verifies the NAS Token and derives the mapped context CK′, IK′ same as in step 1.
6. The MME then transfers the mapped context together with KSI to the SGSN in the Context Response message.
7. On receiving mapped context from the MME, the SGSN replaces any existing UE PS context with the one received.
8. Next the Security Mode Setup procedure of UMTS is run (we do not detail this in this book, the basic idea is same as SMC in EPS).

9. With the security activated, the SGSN sends RAU Accept with P-TMSI and P-TMSI signature to the UE.
10. The UE then responds with a RAU Complete together with optional P-TMSI.

4.17.1.2 UTRAN to E-UTRAN

In E-UTRAN, we talk about TAU instead of RAU. The TAU request and attach request from the UE include the UE security capabilities that the MME stores. Note that the MME does not make use of UE capabilities sent from the SGSN. In the TAU request there is space for the old-GUTI that, together with KSI, is used to identify the UE security context in the MME. A basic message sequence for idle mode mobility from UTRAN to E-UTRAN is shown in Figure 4.23.

Normally the TAU request message will include P-TMSI in the old-GUTI IE (if P-TMSI exists), KSI, old Routing Area Identity (RAI) to identify the SGSN, P-TMSI signature if available, and a 32 bits $NONCE_{UE}$. If the UE has a current EPS NAS security context, then it includes the GUTI and eKSI in the TAU request. The TAU request is integrity protected with the current EPS NAS security context but not confidentiality protected so that the MME can check the GUTI and, if needed, forward the integrity-protected TAU message to the MME with the security context (see Section 4.10.2). It is possible that the current EPS NAS security context is mapped context from earlier key derivation, thereby the eKSI will be KSI_{SGSN}. If a current EPS NAS security context is not available in the UE, the UE sends the TAU request unprotected.

The MME identifies the SGSN with the RAI in the TAU request and sends the context request message together with P-TMSI, P-TMSI Signature, and KSI. The SGSN responds to the MME with the context response message carrying CK || IK. In case the mobility management context in the context response message from the SGSN indicates GSM security mode, the MME will abort the procedure.

If the TAU request was integrity protected then the MME identifies the EPS NAS security context, checks the integrity of the TAU Request message, and derives the AS keys when the active flag in the TAU Request is set or there is pending downlink UP data. The details of behavior regarding TAU Request message is the same as in Section 4.9.2.3.

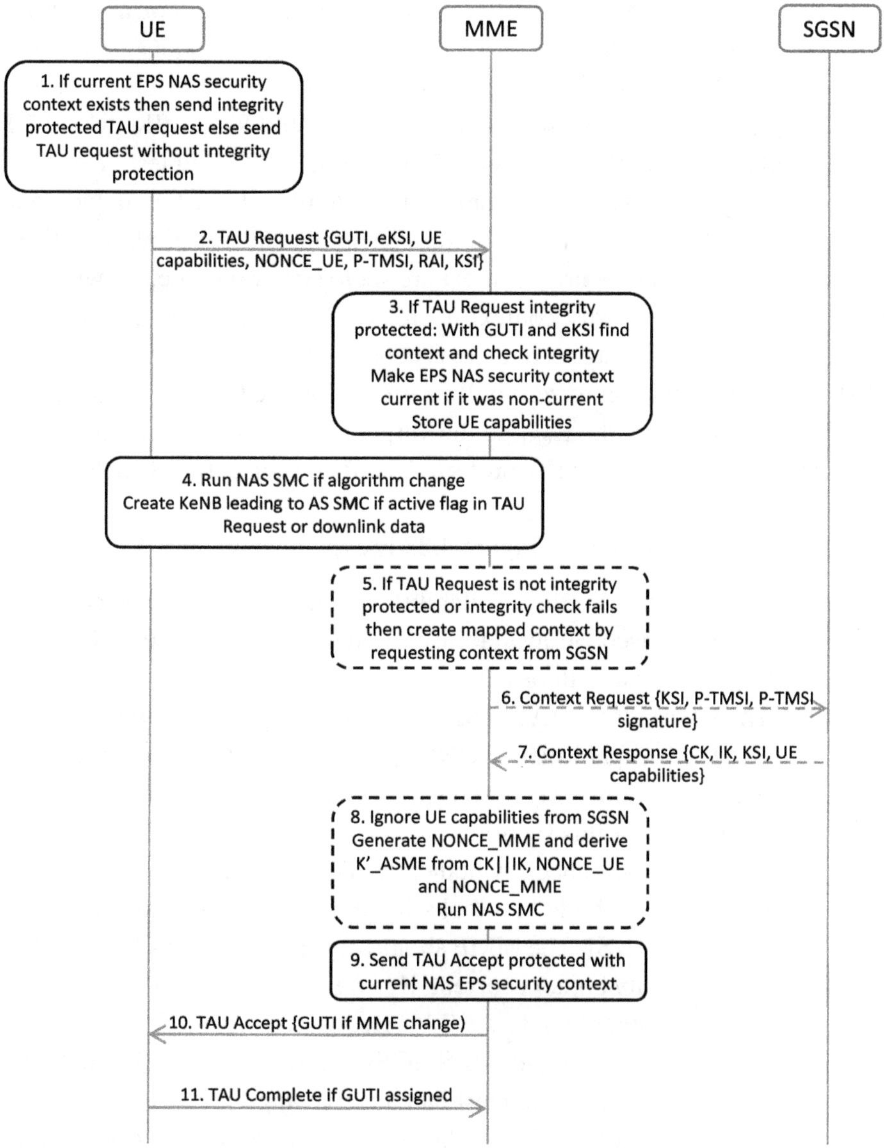

Fig. 4.23 Basic message sequence of idle mode mobility from UTRAN to E-UTRAN.

If the MME does not have the EPS NAS security context indicated by the eKSI from the UE in the TAU request, or the TAU request was received unprotected, the MME will create a new mapped EPS NAS security context (that shall become the current EPS NAS security context). For this purpose the UMTS context is fetched from the SGSN as given above. The MME generates a 32 bit $NONCE_{MME}$ and uses the received $NONCE_{UE}$ with the $NONCE_{MME}$ to derive a fresh mapped K'_{ASME} from CK and IK, where CK, IK were identified by the KSI and P-TMSI in the TAU Request. The MME initiates a NAS Security mode command procedure with the UE including the KSI_{SGSN}, $NONCE_{UE}$, and $NONCE_{MME}$. The uplink and downlink NAS Count for mapped EPS NAS security context are set to start value (i.e., 0) when new mapped EPS NAS security context is created in UE and MME. If the TAU Request had the active flag set or there is pending downlink UP data, the uplink NAS Count, which is set to zero will be used to derive the K_{eNB} in MME and UE. MME will deliver the K_{eNB} to the Target eNB on the S1 interface.

The TAU Accept is security protected using the current EPS NAS security context.

The message sequence of Figure 4.23 is explained below:

1. The UE checks the availability of current EPS NAS context because based on availability of the context the UE will send TAU Request with or without integrity protection.
2. The UE sends the TAU Request message to the MME with GUTI, eKSI, UE capabilities, $NONCE_{UE}$, P-TMSI, RAI, and KSI. Further details on what is sent in TAU Request is as explained in earlier part of this section.
3. Next the MME checks whether the TAU Request is integrity protected, if so, the MME looks for the security context based on GUTI and eKSI. The MME should also make the thus found EPS NAS security context current, if it was non-current, and store the UE capabilities sent in the TAU Request.
4. NAS SMC is run and AS SMC is run only if downlink data is existing or if TAU Request had a active flag.
5. This part happens if the TAU Request is not integrity protected or integrity protection check fails at the MME. In such case the MME creates mapped context based on information from SGSN.
6. Based on reason in 5, the MME sends a Context Request message to SGSN together with KSI, P-TMSI and P-TMSI signature to identify the UE security context in the SGSN and to authorize the UE.

7. As a response to 6, the SGSN sends a Context Response with CK, IK, KSI and UE capabilities.
8. Now the MME generates a $NONCE_{MME}$ that it uses, together with CK, IK and $NONCE_{UE}$, to derive K'_{ASME}. The MME then runs NAS SMC.
9. The MME now prepares the TAU Accept with the NAS EPS security context.
10. The MME sends the protected TAU Accept message to the UE with GUTI if the MME to which the UE got connected to is different.
11. The UE responds to the MME with TAU Complete if GUTI was assigned.

4.17.2 Handover

4.17.2.1 E-UTRAN to UTRAN

During handover from E-UTRAN to UTRAN, the NAS and AS security is activated and thus the UTRAN keys must be derived. The key derivation and related behavior is the same as described in Section 4.17.1.1 for idle mode mobility from E-UTRAN to UTRAN. One important point is that the NAS Downlink Count value is increased by one at the MME when used for key derivation so as to prevent its reuse in future to create mapped context. As the MME does not send any NAS message to UE during handover, the 4 LSBs of the NAS Downlink Count is sent to the UE in the handover command message. The UE estimates the NAS Downlink Count with the 4 LSBs and stores the new NAS Downlink Count value if it is higher than the one the UE already has; as discussed earlier, the NAS Count is never decreased.

MME transfers the UE security capabilities and the mapped context to the SGSN. UMTS network then selects the algorithm and activates appropriate security.

If the handover is not completed successfully, then the mapped UMTS security context cannot be used in the future. The SGSN will delete the mapped UMTS security context and the stored UMTS security context, which has the same KSI as the mapped UMTS security context.

After HO from E-UTRAN to UTRAN, it is optional for EPS to keep the current EPS NAS security context. If the EPS NAS security context is kept, then it is stored as the current EPS NAS security context in E-UTRAN in the UE and the MME.

4.17.2.2 UTRAN to E-UTRAN

During handover, service continuity without perception of service quality degradation by the user is the most important point. Therefore any system design must try to minimize delay due to handover. Minimizing of delay during handover must go hand-in-hand with maintenance of security level, i.e., handover should not lead to degraded security. Thus for EPS handover from UTRAN to E-UTRAN it is necessary that the security association, keys, and algorithms to be used are setup before the UE starts accessing service in E-UTRAN. The solution is to take the E-UTRAN keys in use at the time of handover or to create mapped security context from the security context being used in UTRAN. Obviously it is possible that the given UE or MME do not have the native EPS security context thus requiring negotiation messages for which there is no time during handover at the same time handover does the native EPS security context, if available, will be stored in the form of NAS security context requiring NAS messages for activation. Thus the obvious solution is to create and use mapped security context at handover from UTRAN to E-UTRAN, i.e., UTRAN security context mapped to E-UTRAN security context. Native EPS NAS security context in UE and MME can be taken in use at a later time.

On handover the algorithm selected is the one of preference at the target network. The NAS security context is activated at the UE by handover command message, this also leads to activation/derivation of the AS keys while the NAS keys are activated in MME by the handover notify message and the AS keys are activated in the eNB by the handover complete message. The security context used for the handover is the most recent context being used at the UTRAN, be it generated by UMTS AKA or a mapped context from previous handover from E-UTRAN to UTRAN. The message sequence is shown in Figure 4.24.

As a first step, the source RNC makes a decision regarding handover such as to prevent ping-pong handover between UTRAN and E-UTRAN. Once a handover decision is made, the RNC sends a Relocation Request message to the source SGSN. As a result the SGSN transfers the CK, IK (or Kc), KSI, and

UE security capabilities to the chosen MME together with the UE EPS security capabilities. Based on the information received the MME will decide the algorithm to be used for the given UE. In case that the source SGSN is a Release 7 or earlier SGSN then it will not understand the UE EPS security capabilities sent by the UE during Attach Request or RAU Request. Thus the SGSN will not send any UE EPS security capabilities to the MME. In such case the MME will choose one of the default algorithms, as per operator preference given by the algorithm priority list, for use with the UE. Having chosen the algorithm the MME will derive K'_{ASME} using the concatenated CK and IK together with a $NONCE_{MME}$ that will be associated with the KSI_{SGSN}. The order of choosing algorithm and key derivation can be implementation-specific. Next the MME will derive the NAS keys and K_{eNB} from K'_{ASME}. The NAS Count values will be set to zero (also known as start value) in both MME and UE.

As NAS messages are not sent during handover, the MME sends the chosen algorithm, KSI_{SGSN} and $NONCE_{MME}$ to the Target eNB in a so called NAS transparent container IE together with the S1 handover request message. The S1 handover request message also includes the K_{eNB} and the UE security capabilities, received from SGSN or default algorithms. The Target eNB then chooses an algorithm for AS security and creates another transparent container, known as RRCConnectionReconfiguration, (Handover Command) which includes the transparent container from the MME and chosen AS algorithms. This, unprotected, RRCConnectionReconfiguration transparent container is sent to the MME in the S1 handover request ACK message. The MME forwards the thus created transparent container to the SGSN in Forward (FW) Relocation Response message that is further sent to the RNC in the relocation command. The RNC sends the transparent container to the UE in the UTRAN handover command, which is integrity protected and optionally confidentiality protected. The UE then derives the K'_{ASME}, and the NAS and AS keys, and sends the handover complete message in the form of RRCConnectionReconfiguration Complete message to the Target eNB (now the eNB to which the UE is attached to). At this time, the NAS Uplink and Downlink Counts for the mapped context at the MME and UE are set to zero. The handover complete message and the following AS messages are confidentiality and integrity protected as decided during previous messages based on operator policy. The mapped context now becomes the current EPS security context and any native EPS security context becomes non-current EPS security context overwriting any existing non-current context. The thus created mapped context is used only once that means that the context is removed in case of unsuccessful handover.

It is recommended that the EPS security context is used when handover happens to E-UTRAN so as to avoid any issues arising from successful attack that might have happened in UTRAN. The means to change the security context to mapped security context is the TAU Request message from the UE. The behaviour of the system regarding TAU Request message is similar to that in idle mode mobility and state change. The TAU Request message is integrity protected by the mapped context and contains KSI_{SGSN}, KSI_{ASME} of the native context together with GUTI and UE security capabilities. The MME checks the UE security capabilities and verifies whether it is the same as that received from the SGSN. In case the UE security capabilities are different, then the MME verifies whether the highest priority algorithm is being used for the UE else the MME runs the NAS SMC and also informs the eNB of the correct UE EPS security capabilities that in turn triggers a change in AS algorithm if required. The rules regarding the TAU request message are (also discussed earlier in the chapter):

If a TAU is not received for a specified time after handover then the MME considers that the UE shares the same native EPS NAS security context. This because the TAU request is conditionally sent, e.g., when the TA changes therefore the UE may not send a TAU request.

When the TAU request contains KSI_{ASME} and GUTI, the MME may optionally take the given native EPS security context, if available, as the current context and remove the mapped context. This means a NAS SMC, K_{eNB} key change on the fly and assignment of a new GUTI if the TAU Request GUTI pointed to a different MME. Other ways to change the keys are (1) transition of ECM-IDLE state and associated behavior or (2) transition to EMM-DEREGISTERED state.

On the other hand, if the MME does not receive KSI_{ASME} and GUTI then it removes the non-current native EPS security context.

The message sequence of handover from UTRAN to E-UTRN is given in Figure 4.24 and explained below:

1. After handover decision is taken the Source RNC sends a Relocation Request to SGSN.

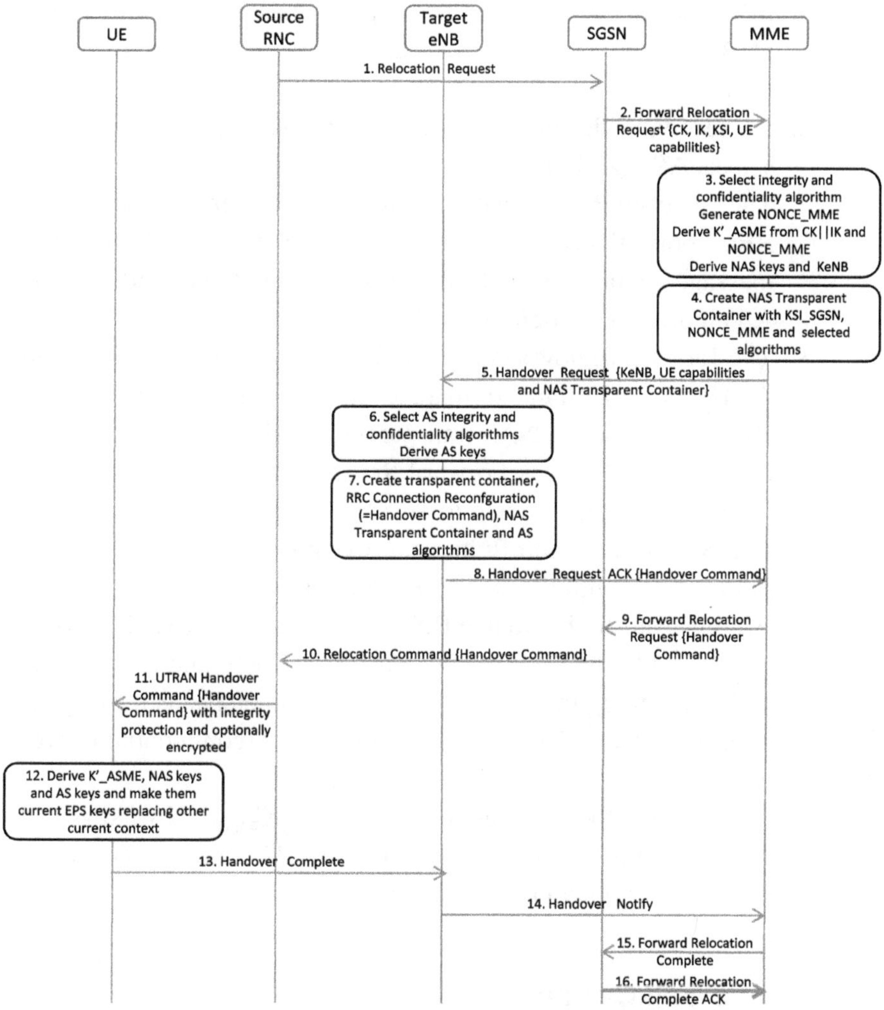

Fig. 4.24 Handover from UTRAN to E-UTRAN.

2. The SGSN in turn sends Forward Relocation Request to MME with CK, IK, KSI and UE capabilities.
3. On receiving the UMTS context, the MME now chooses the integrity and confidentiality algorithms, and the MME derives mapped K'_{ASME} using $NONCE_{MME}$ and concatenated CK, IK.
4. The MME also creates a NAS transparent container, meant for the UE, with KSI_{SGSN}, $NONCE_{MME}$, and MME selected algorithms.

5. MME sends the Handover Request message to the Target eNB together with K_{eNB}, UE capabilities, and NAS transparent container.
6. On receiving the information from the MME, the Target eNB selects AS algorithms and derives AS keys.
7. The Target eNB also creates a transparent container in the form of RRCConnectionReconfiguration (i.e., the Handover Command) message containing the NAS transparent container and also the AS algorithms chosen by the Target eNB.
8. The Target eNB now sends a Handover Request to the MME with the Handover Command (that has the transparent container from Target eNB and the NAS transparent container).
9. Now the MME sends Forward Relocation Request (see step 2) with Handover Command to SGSN.
10. The SGSN sends the Relocation Command (see step 1) to the Source RNC with the Handover Command.
11. All this leads to the Source RNC sending the UTRAN Handover Command to the UE with the Handover Command from Target eNB containing the two transparent containers. This message is integrity protected and optionally encrypted (confidentiality protected).
12. On receiving the message the UE derives K'_{ASME}, NAS keys and AS keys. This key set becomes current EPS security context and replaces any other current context.

4.18 Voice Call Continuity

Obviously any new network deployment will not mean complete coverage from day one. Therefore an operator will have to depend on provisioning of services over its existing network. One of the important services for any operator is voice services. EPS is packet switched solution therefore voice service is provided over IP multimedia subsystem (IMS) in the form of voice over IP (VoIP). This means that a given deployment by the operator might require handover of voice service to circuit switched legacy network of the operator. This is known as single radio voice call continuity (SRVCC) and is specified in 3GPP TS 23.216. For this purpose the mobile switching center (MSC) server needs to be enhanced to handle such calls. The security part works similar to that of PS handover from E-UTRAN to UTRAN except that the derived keys are known as CK_{SRVCC} and IK_{SRVCC} instead of CK' and IK'. If the legacy network of the operator supports PS services then obviously non-voice services in EPS can also be handedover. The whole procedure is same as in Section 4.17.2.1.

4.19 Emergency Call Security

Supporting emergency calls is mandatory in almost all countries therefore emergency call security solution is developed by 3GPP. A UE can make emergency call at anytime, i.e., be it authenticated or not. At the same time the network can be configured to accept unauthenticated emergency calls or only authenticated emergency calls. A UE that is not capable of authenticating to the network due to unknown USIM to the network or absence of USIM is known to be in limited service mode (LSM).

A UE that is capable of performing authentication will try to do so and thus provide authenticated emergency call service. If by any reason the authentication fails, the UE will try to attach to the network by using Attach Request message with type Emergency or simply known as Emergency Attach. In the Emergency Attach the UE sends GUTI or P-TMSI if available, else the UE will send the IMSI and if none of these are available then the UE will include the IMEI. On receiving an Emergency Attach the MME does not send Notify Request to HSS for AV. EIA0 and EEA0 are used as algorithm for cases where unauthenticated emergency call is provided. The same is valid for a UE in LSM.

Bibliography

Accessing the specifications:

- To find all documents of specific series: http://www.3gpp.org/ftp/Specs/html-info/xx-series.htm. Replace xx by series, e.g., 33 for 3GPP security specifications, 36 for RAN specifications, 24 for NAS specifications and 23 for stage 3 specifications.

- To find a given specification: http://www.3gpp.org/ftp/Specs/html-info/xxyyy.htm. Replace xxyyy by specification number, e.g., 33401 for TS 33.401.

[1] 3GPP TR 21.905, Vocabulary for 3GPP Specifications.
[2] 3GPP TS 23.401, General Packet Radio Service (GPRS) enhancements for Evolved Universal Terrestrial Radio Access Network (E-UTRAN) access.
[3] 3GPP TS 23.003, Numbering, addressing and identification.
[4] 3GPP TS 33.102, 3G security; Security architecture.
[5] 3GPP TS 33.210, 3G security; Network Domain Security (NDS); IP network layer security.
[6] 3GPP TS 33.310, Network Domain Security (NDS); Authentication Framework (AF).
[7] 3GPP TS 24.301, Non-Access-Stratum (NAS) protocol for Evolved Packet System (EPS); Stage 3.
[8] 3GPP TS 36.323, Evolved Universal Terrestrial Radio Access (E-UTRA); Packet Data Convergence Protocol (PDCP) specification.
[9] 3GPP TS 31.102, Characteristics of the Universal Subscriber Identity Module (USIM) application.
[10] 3GPP TS 35.215, Confidentiality and Integrity Algorithms UEA2 & UIA2; Document 1: UEA2 and UIA2 specifications.
[11] NIST, Advanced Encryption Standard (AES) (FIPS PUB 197).
[12] NIST Special Publication 800-38A (2001), Recommendation for Block Cipher Modes of Operation.
[13] NIST Special Publication 800-38B (2001), Recommendation for Block Cipher Modes of Operation: The CMAC Mode for Authentication.
[14] 3GPP TS 36.331, Evolved Universal Terrestrial Radio Access (E-UTRA) Radio Resource Control (RRC); Protocol specification.
[15] 3GPP TS 23.216, Single Radio Voice Call Continuity (SRVCC); Stage 2.
[16] ZUC Discussion Forum, http://zucalg.forumotion.net/.
[17] GSMA Security Algorithms Web site, http://www.gsmworld.com/our-work/programmes-and-initiatives/fraud-and-security/gsm_security_algorithms.htm.
[18] NTT DOCOMO Technical Journal, Volume 11, No. 3, September 2009.

5

Security in IEEE 802.16e/WiMAX

In this chapter, we present the security features of WiMAX. We first discuss briefly the architecture and operation of the WiMAX security systems and functions. Next, we describe the details of the WiMAX security system in terms of encryption, authentication, and key management. Specifically, we cover the subjects on WiMAX security requirements, i.e., privacy key management (PKM), authentication and authorization, preauthentication, traffic encryption key (TEK) updates, etc. with the focus on the security sublayer in the MAC layer.

5.1 Overview

5.1.1 WiMAX Protocol Stack

We start this chapter by first reviewing the structure of the lower two layers of WiMAX. The protocol architecture of WiMAX is structured into two main layers: the medium access control (MAC) layer and physical layer as can be seen in Figure 5.1. The diagram also indicates interfacing points where service access points (SAPs) are formally defined by the standard. Basically, connections are established in advance, and each slot belongs to a certain connection identified by a connection identifier (CID). The concepts of connection and CID are defined as follows [14]:

- A connection is a unidirectional virtual mapping between a BS and an MS MAC peer for transporting a service flow's traffic, and is

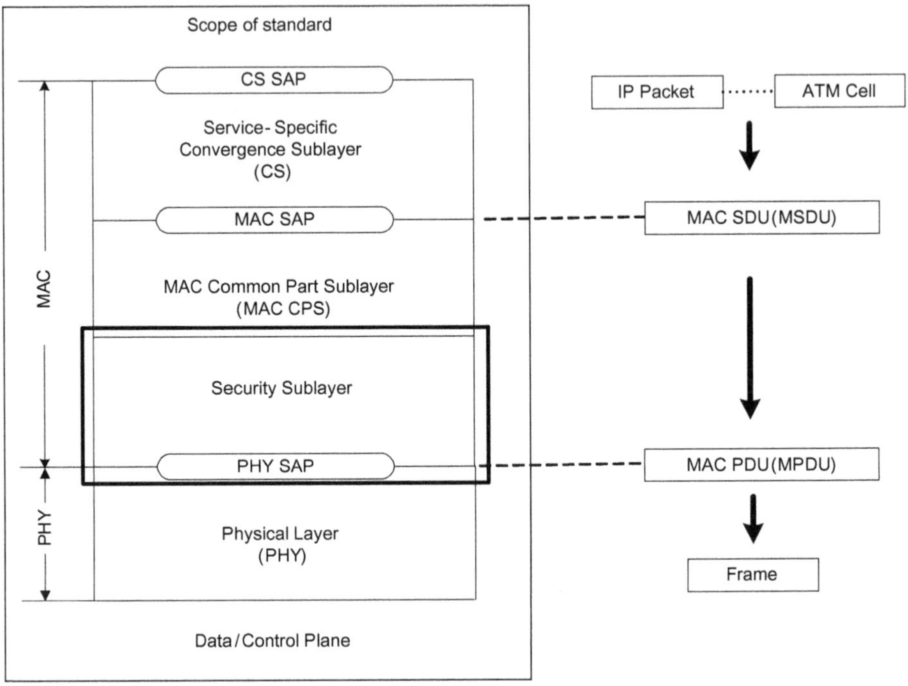

Fig. 5.1 WiMAX protocol stack.

defined only for one type of service. Typically, there are two types of connections: transport connection and management connection.

- A CID identifies a MAC level connection between a BS and an MS. A CID is 16-bit value.

The central element of the layered architecture is the MAC common part sublayer (MAC CPS). In MAC CPS, MAC protocol data units (MPDUs) are constructed, connections are established, and bandwidth is managed. The MAC CPS exchanges MAC service data units (MSDUs) with the service-specific convergence sublayer (CS). The CS maps the external units of data (e.g., IP packets or ATM cells) of higher level protocols into the MSDU format. The CS also classifies the MSDUs and associates them with the proper MAC connection identifier (CID). Classifying and mapping the MSDUs into appropriate CIDs are the basic functions of QoS mechanism of WiMAX.

The MAC CPS is tightly integrated with the security sublayer, which must be included to protect the information confidentiality and integrity

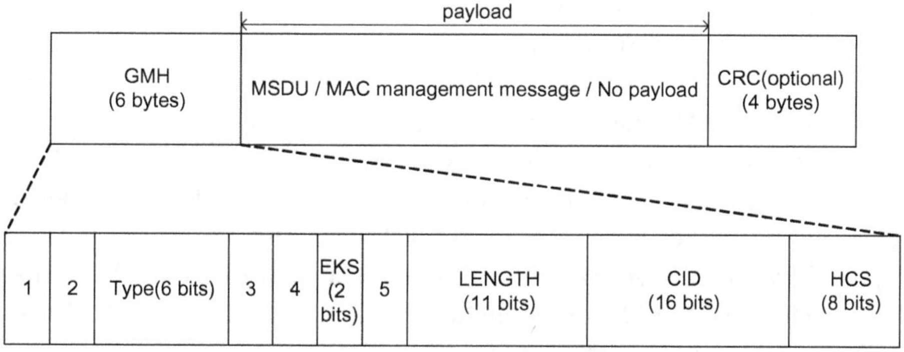

Fig. 5.2 MPDU format with GMH.

considering the open nature of wireless channels in the physical layer (PHY). The PHY plays a role of two-way mapping between MPDUs and PHY frames received and transmitted through coding and modulation of radiofrequency (RF) signals.

The MAC PDU has the format as shown in Figure 5.2. The typical format of the MPDU is composed of three fields, i.e., header, MSDU, and cyclic redundancy checking (CRC). MPDUs can be differentiated in two forms by the header:

- Generic MAC header (GMH): a payload and an optional CRC follow
- Bandwidth request header (BRH): the header becomes an entire packet.

Depending on the type of CID within the header, the MPDU frame is used either for transporting the higher layer data packets or for delivering the management messages. Transport connections carry MSDUs that are passed by the network stacks above MAC. In management connections, each management MPDU carries a single MAC management message. Management connections take care of broadcast data, initial ranging, bandwidth requests, and general management messaging. Thus, WiMAX affords much flexibility as to how MPDUs carry MSDUs.

As for management connections, WiMAX defines three types: basic, primary, and secondary. The basic management connection is used by the BS MAC and MS MAC to exchange short and urgent MAC management messages while the primary connection is used to exchange longer and more delay-tolerant MAC management messages. The secondary management connection is used to transfer delay tolerant and higher-layer IP management datagrams, which are not MAC management messages. All other connections are transport connections. Usually, when an MS enters a network and is initialized, two management connections, one for uplink and another for downlink connections, are established between MS and BS. The third may be established in an option base. (Note that IEEE 802.16e denotes a mobile user as a subscriber station (SS), while WiMAX Forum does as a mobile station (MS). We use them interchangeably without confusion.)

Among the connections, WiMAX protects only transport connections and the secondary management connection (i.e., the basic and primary connections are not secured). To establish secure connections, MS and BS use the encryption-related information that is contained in the *generic MAC header* (GMH) as shown in Table 5.1. The contained information, encryption control (EC) and encryption key sequence (EKS), is used to decrypt the encrypted payload at the receiver. Specifically, the EC field indicates whether the payload carried by the GMH is encrypted or not: EC = 1 indicates that the payload is encrypted, and EC = 0 not encrypted. The two bits in the EKS field delivers the key sequence number (or index) of the TEK, and is meaningful only when EC = 1.

Table 5.1. Field description of the generic MAC header (from [1]).

Field Name	Field # in Fig. 2	Description
HT	1	Header type 0 for generic MAC header
EC	2	Encryption Control: 0 = payload is not encrypted; 1 = payload is encrypted. For a MAC header without payload, this bit indicates the frame type
Type		Indicates the subheader and special payload type
ESF	3	Extended Subheader Field
CI	4	CRC Indicator: 1 = CRC included; 0 = no CRC included
EKS		Encryption Key Sequence. The index of the TEK and initialization vector used to encrypt the payload. This field is meaningful only if the EC field is set to one
Reserved	5	
LEN		Length of MPDU in bytes
CID		Connection IDentifier
HCS		Header Check Sequence

5.1.2 Architecture of WiMAX Security Sublayer

The security sublayer in WiMAX MAC layer provides core security functions to MPDUs exchanged with PHY as shown in Figure 5.1. The functions cover device and/or user authentication, protection of integrity of control and management messages, and protection of privacy of message contents, etc. For this purpose, WiMAX has included diverse security protocols and database servers for authentication, authorization, and key management, etc. Basically, WiMAX defines two security component protocols: an encapsulation protocol and a secure key management protocol.

To provide privacy, the encapsulation protocol defines a set of supported cryptographic suites (i.e., pairings of data encryption and authentication algorithms) and the rules for applying those algorithms to an MPDU payload. By the encapsulation protocol, the data transmission between MS and BS is encrypted to keep it unavailable to other principals. It is noted that only the MPDU payload part is encrypted while the generic MAC header is not encrypted, which is to facilitate registration, ranging, and normal operation of MAC. It is also noted that the control and management messages are under the protection of integrity so that an adversary shall not be able to modify these messages. In the message integrity aspect, the security mechanisms support the functions of message authentication code (MAC) that guarantee both message integrity and source confirmation. This mechanism also resists from the replay attack.

The key management protocol, PKM, provides the secure distribution of authorization and encryption keying materials between BS and MS. Through PKM, it is possible for MS to share the same keying data with BS and is also possible for BS to enforce conditional access to network services. The MS uses the PKM protocol to obtain authorization and keying material from the BS, and to support periodic reauthorization and key refresh. The PKM protocol allows for both mutual and unilateral authentication between the BS and the MS.

Authentication is a basic function required when the MS verifies itself with the BS at network entry. This initial authentication must be strong in order to prevent an illegal MS from entering the network. In the WiMAX network, the BS should identify the MS being served serving through device authentication. When different users (real human or an application) share a platform on one mobile station or terminal, user authentication may also be needed. Note that the authentication between the subscriber and the 802.16 system may be handled by higher layer protocols, which is beyond the scope of the specification.

In WiMAX, authentication is done based on either RSA or extensible authentication protocol (EAP). For authentication purpose, PKM uses one of the three methods: (1) X.509 digital certificates together with RSA public key encryption algorithm; (2) EAP-based authentication; (3) a sequence starting with RSA-based (or EAP-based) authentication and followed by EAP authentication. The dashed block on the right top in the Figure 5.3 indicates that the EAP-based authentication is done by the EAP methods between the AAA server and the MS. The related EAP messages are encapsulated within the MAC security sublayer frames. Supporting the RSA-based protocol is mandatory in PKMv1 but is optional in PKMv2. On the other hand, supporting the

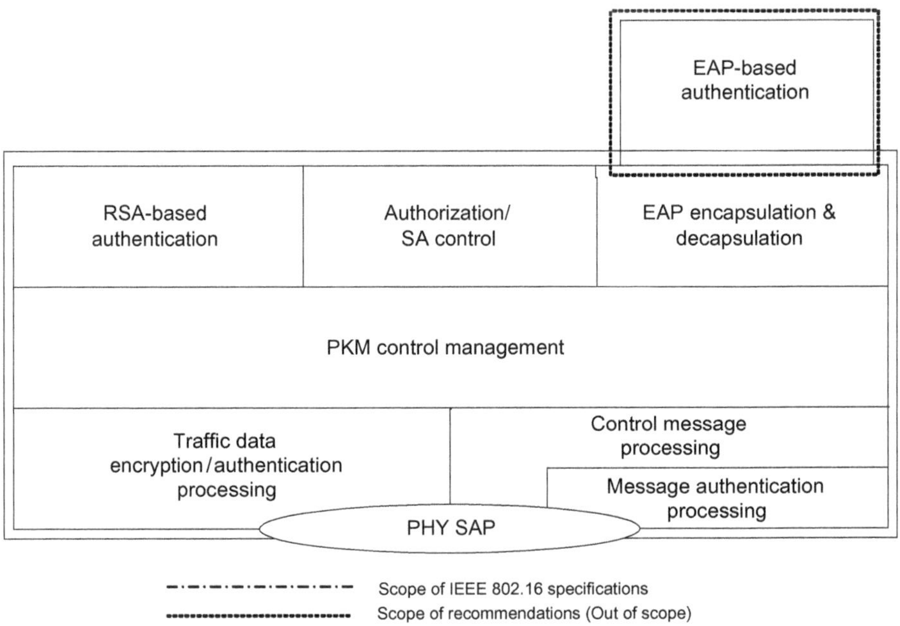

Fig. 5.3 Components of security sublayer (see [1]).[1]

[1] IEEE Std 802.16e — IEEE Standard for the Software reviews ©2005 IEEE. All rights reserved.

EAP-based protocol is optional unless specifically required. It is noted that for the user-level authentication, EAP-based and EAP-SIM (Subscriber ID Module)-based mechanisms can be utilized.

Authorization is a security process required to determine what services an authenticated subscriber is permitted to invoke. Each subscriber has a set of credentials that describe what the subscriber is allowed to do. For requesting authorization in WiMAX, an MS first presents its credentials to the BS, and then the BS authenticates the MS by using the credentials during the initial authorization exchange. The credentials are a unique X.509 digital certificate issued by the manufacturer in the case of RSA authentication and an operator-specified credential in the case of EAP-based authentication. The authorization process is performed for each type of authentication method, and generates a shared secret (called an authorization key (AK)) to authorize the MS.

Authorization is followed by the PKM control process, which eventually generates the TEK from AK and passes it to the MS. Once the TEK is shared by the BS and the MS successfully, the traffic data encryption and authentication process follow in the data plane. The basic concept of the two-tiered mechanism for key distribution is refreshing the TEKs without incurring the overheads of computation-intensive operations. Control message processing and message authentication process take place in the control plane. The PKM protocols support periodic reauthentication/reauthorization and key refresh.

Through authorization with the AK exchange, the BS can confirm whether an MS's authenticated identity corresponds to a valid subscriber, and hence he/she is authorized to access the data services corresponding to specific TEKs. In this way, the BS may be protected against an attacker masquerading a legitimate subscriber by employing a cloned MS.

5.1.2.1 PKM RSA Authentication

The PKM RSA authentication protocol uses X.509 digital certificates and the RSA public key encryption algorithm. The protocol binds the RSA encryption public key and the MAC address of the MS, which are contained in the digital certificate. When requesting an AK to the BS, the MS presents its digital certificate together with its cryptographic capability and its CID. The BS verifies the digital certificate; then uses the verified public key to encrypt an AK and sends back the AK to the requesting MS.

All the MSs that rely on RSA authentication must have manufacturer-issued RSA private/public key pairs. Or, they have to provide an internal algorithm to generate such key pairs dynamically prior to the first key exchange. All the MSs with factory-installed RSA key pairs must also have the manufacturer-issued X.509 digital certificates. In the case of using internal algorithms, the MS must support a mechanism for installing a manufacturer-issued X.509 certificate following key generation.

5.1.2.2 PKM EAP Authentication

PKM EAP authentication uses EAP in conjunction with an operator-selected EAP method, for example EAP-TLS (transport layer security), which uses a particular type of credential, i.e., the X.509 digital certificate in the case of EAP-TLS or a subscriber identity module (SIM) in the case of EAP-SIM. The particular credentials and EAP methods that are to be used are out of the scope of IEEE 802.16e standard, but the selected method must fulfill the "mandatory criteria" listed in RFC 4017.

The EAP transfer messages that are delivered to the BS are protected in two ways. At initial authorization, the messages are protected with EAP integrity key (EIK), while during reauthorization, they are protected with an HMAC/CMAC tuple. Thus, when valid EAP transfer messages are received by the BS, they are immediately sent to the AAA server via Diameter protocol. However, when any EAP transfer messages with invalid EIK during initial authorization or any messages with invalid HMAC/CMAC Digests during reauthorization are received, they must be discarded.

5.1.2.3 Authentication in Network Entry Procedure

When a node enters a WiMAX network, systems must follow a series of procedures for registering the new MS as shown in Figure 5.4. In this section, we briefly describe the overall network entry procedure only for PMP (point-to-multipoint) topology, and explain at which point the MS authentication and key exchange occur.

1. The MS scans for downlink channel signals and synchronizes with the BS. Channel parameters are derived from scanning.

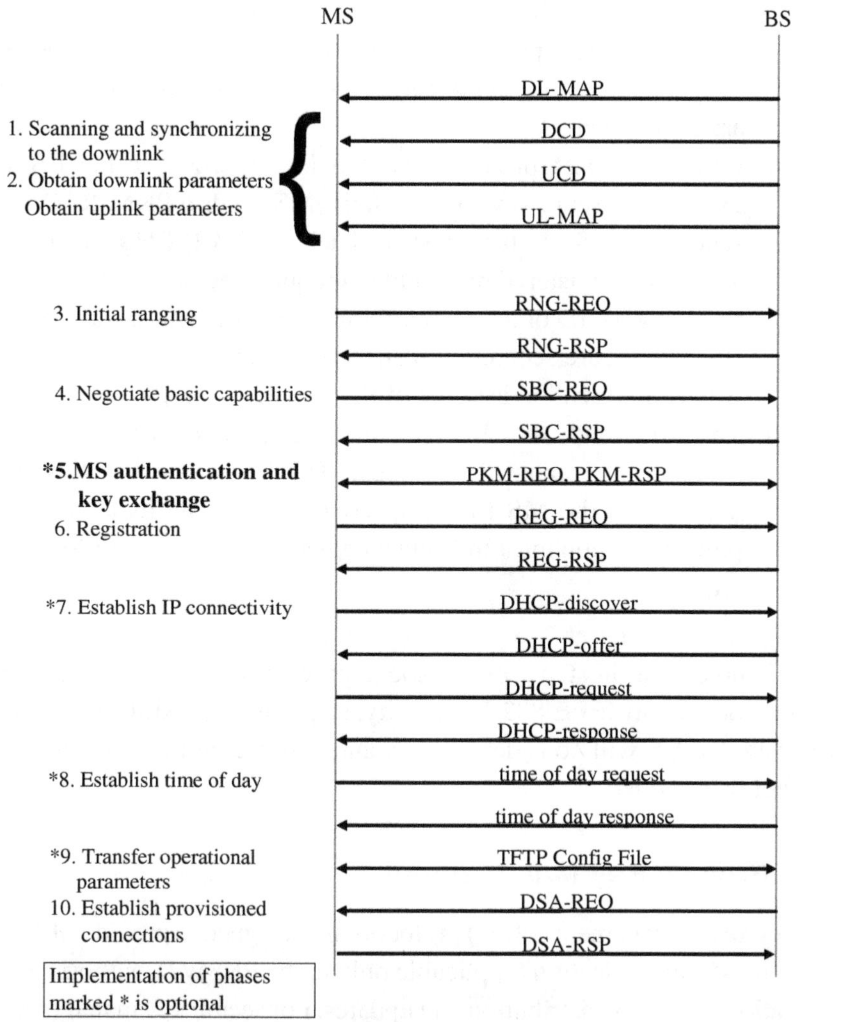

Fig. 5.4 Network entry procedures.

2. The MS performs initial ranging to adjust PHY parameters, i.e., timing offset and power parameters, and establish the basic and primary unsecure management connections by sending an RNG-REQ message to the BS in a contention-based initial ranging interval. The primary management connection is used for capability negotiation, authorization, and key management.

3. The MS notifies the BS of its basic capabilities by sending an SBC-REQ (SS basic capability request) message. Through this message, the parameters relevant to bandwidth allocation and physical layer are exchanged.
4. Using the PKM protocol, the MS is authorized to the BS and exchanges secure keys. For this, the MS sends a PKM-REQ message to the BS and the BS sends back a PKM-RSP as a response.
5. The MS is registered by sending a request message to the BS and receiving a CID of a secondary management connection from the BS. The secure secondary management connection is established when authorization is successfully finished during registration.
6. The MS and the BS set up transport connections using MAC_create_connection request. The BS sends the DSA-REQ message to the MS to set up connections. The dynamic transport connection must indicate whether the encryption of MPDU is required.

It is noted that in step 3 of the above procedure, if the MS specifies that it does not support IEEE 802.16 security, step 4 is skipped. If provisioned in this way, the BS will consider the MS authenticated; otherwise, the service will not be provided.

5.2 Privacy Key Management

Privacy key management (PKM) protocols are designed to provide MS authentication, BS authentication (applicable only in the PKMv2), authorization key derivation, and TEK distribution and updates. For secure key management and agreement, the PKM protocols meet the following requirements. First, more than one credential should be provided in the authentication process. Second, the key management and agreement protocol should be modern and publicly analyzed and confirmed. Third, keys should be fresh and updated within a lifetime. Fourth, the perfect forward secrecy is recommended to prevent an old session key from being reused in case a long-term key is compromised. Fifth, an efficient key management mechanism such as a hierarchical key management scheme should be devised. Each of the keys should be tightly bound to the authentic identity of a principal to prevent the masquerading attack. Finally, the key management and agreement protocol should guarantee a secure key translation or transport between MAC layer principals.

In the following, PKMv1 and v2 are specified in detail, and a comparison between them is given. We first start by introducing the concept of security association in WiMAX. The keys and associations established between an MS and a BS during the authorization step at network entry are discussed in the sequel.

5.2.1 Security Associations

A security association (SA) is the set of security information a BS and one or more of its client MSs share in order to protect connections across the WiMAX network. Through an SA, the BS and the MS set the security parameters of a connection, e.g., keys and selected encryption algorithms with initialization vectors. When transmitting and receiving an MPDU, the sender/receiver perform encryption/decryption and data authentication on the MPDU payload as specified by the SA. Thus, if an MPDU received on a connection mapped to an SA that requires encryption turns out not to be encrypted, it is discarded.

In [15], the authors classify the SAs into two types: data SA and authorization SA. The BS uses authorization SA to configure data SA on the MS. The exact content of the SA depends on the SA's cryptographic suite, but the general contents of the two SA's can be listed as in Table 5.2.

The core elements of an authorization-related SA are X.509 certificate, AK, key encryption key (KEK), and HMAC key. A KEK, a 112-bit Triple-DES

Table 5.2. Contents of SA.

Contents of data-related SA	Contents of authorization-related SA
• A 16-bit SA identifier (SAID), which is unique within a BS • Cipher algorithms • Two 128-bit TEKs • Two 2-bit key identifiers • TEK lifetime (default is half a day; minimum is 30 min; maximum is 7 days) • A 64-bit initialization vector for each TEK in DES-CBC mode • Type indication of SA (primary, static or dynamic)	• A X.509 certificate for MS • A 160-bit AK • A 4-bit AK identifier • AK lifetime (default is 7 days; minimum is one day; maximum is 70 days) • A 128-bit KEK • A downlink HMAC key • An uplink HMAC key • A list of authorized data SAs

key, is used for distributing TEKs. A hash function-based HMAC key is for providing data authenticity of key distribution messages between the BS and the MS. A list of authorized data SAs is managed by SA descriptors, each including the SAID, the SA type, and the SA cipher suite.

There are three types of SAs, namely, primary, static, and dynamic. There is one primary SA for each MS. The primary SA is established when the MS is initialized and is used exclusively between an MS and its BS. Static SAs are created on the BS during the initialization of an MS. The static SA is used for the basic unicast service. If an MS has subscribed to additional services, there are as many additional static SAs as the additional services. Dynamic SAs are established and eliminated dynamically in response to the initiation and termination of specific service flows. Static SAs and dynamic SAs can be shared among many CIDs when multicast is used. Each SA is identified using its unique 16-bit SAID within a BS. For each MS, the SAID of its primary SA is equal to the basic CID of the MS.

As for the connections, the basic and primary management connections do not have SAs although the integrity of management messages can be secured. On the other hand, an SA for the secondary management connection is automatically created on an optional basis on network entry. Transport connections always have SAs. Thus, an MS typically has two or three SAs, one for the secondary management connection and either one for both uplink and downlink transport connection or separate SAs for uplink and downlink connections.

5.2.2 PKMv1

5.2.2.1 Overview of PKMv1

Using the PKM protocol, an MS requests its keying material to its BS. But, before doing that, the BS must make sure that each MS has access only to the SAs it is authorized to access. Every MS is preconfigured with an X.509 certificate, which contains the public key (PK) of the MS. IEEE802.16 standard utilizes two types of certificates: manufacturer's and MS's. The manufacturer's certificate is signed by itself or the third party. The MS's certificate is created and signed by its manufacturer. The MS uses it for its authentication with the BS. All other keys are established during or after authorization.

Once the BS determines the AK, the BS encrypts it using the PK and passes it to the MS. The AK has a sequence number (from zero to 15) and a lifetime. The MS uses the AK to determine the KEK and HMAC key. The sequence number of the AK implicitly belongs to the HMAC keys as well. KEKs are used to encrypt TEKs during their transfer. Since the keying material associated with an SA has a limited lifetime, BS refreshes the SA's keying material periodically through reauthorization. The lifetime of an AK ranges from 1 to 70 days, with a default value of 7 days. The lifetime is a 32-bit unsigned number representing the number of seconds until AK expires. In order for the smooth transition of keys in case the lifetime of keys expires, two AKs may be simultaneously active with overlapping lifetime in each SA. For this purpose, the MS maintains two most recent generations of keying materials for each SA.

Two versions of keying materials are identified by the 2-bit key identifier, which is distributed to the MS. The BS and the MS use this identifier to make sure whether they are synchronized in using the same keying materials to encrypt the MPDU payload. The key identifier is updated by the BS (automatically wraps around to 0 from 3) whenever the BS generates new keying materials.

5.2.2.2 Detailed Procedure of PKMv1

There are basically two phases in the PKM. In PKMv1, they are respectively the authentication/authorization phase and the TEK request/update phase. The first one is the PKM authorization protocol, which is established by the MS and responded by the BS. At the end of a successful run of this protocol, an AK is created by BS and transmitted to MS. After that, each party generates a KEK, and eventually a TEK. TEKs can be taken as session keys, while AK/KEKs are long-term keys. Then, in the second part, the PKM protocol allows MS to gather TEKs from BS.

The flow of the protocols in PKMv1 is shown in Figure 5.5. In the figure, the PKM protocol relies on two generic MAC management messages defined in the 802.16 standard. The first one is the PKM Request (PKM-REQ) and the second one is the PKM Response (PKM-RSP). These messages encapsulate one PKM protocol message in its message payload among authorization information, authorization request, authorization response, key request, or key reply. Both the messages use the primary management connection in a unicast service.

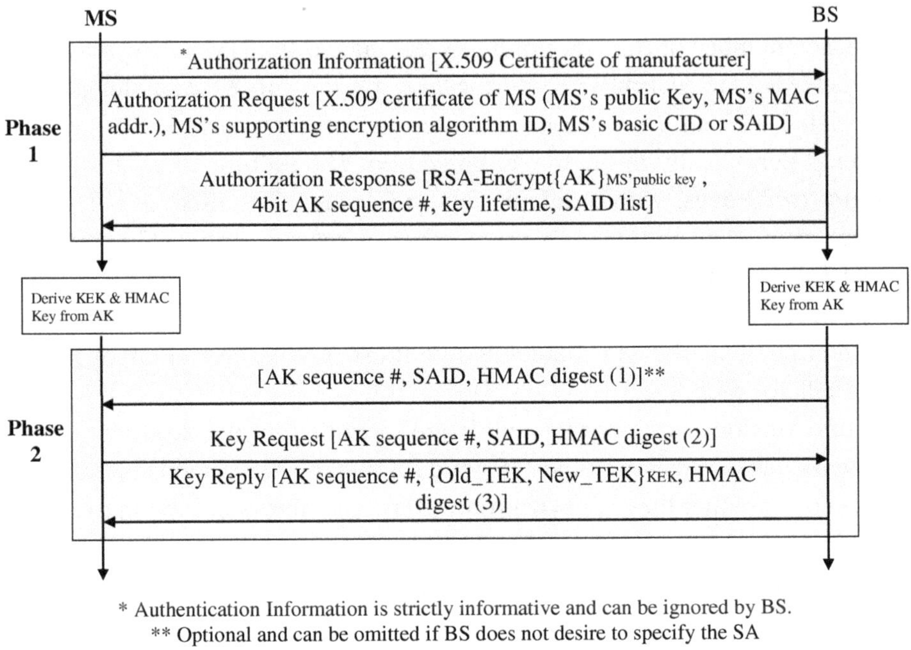

Fig. 5.5 Message flow of PKMv1.

The PKM authorization protocol is for exchanging three messages between an MS and a BS. In the first message, the MS sends the BS the certificate of its manufacturer. Soon after the first message, the MS sends the second message, the authorization request, including its own certificate, capabilities, and an SAID to BS. SAID maps to the SA that MS wishes to adopt during the following protected procedure. After the BS authenticates the MS by verifying the two messages, the BS replies to the MS with an authentication message, within which AK, encrypted by the MS' public key, lifetime of AK, sequence number of AK, and a list of SAID for MS' future selection of security associations are contained.

In the second phase, after the authentication and authorization process is completed, the MS requests TEK to the BS, or updates TEKs. In this phase, a data SA is established between the BS and the MS. Of the three messages, the first message can be omitted, if the BS does not desire to rekey a data SA or create a new SA. In the second message, the MS sends the TEK request to the BS, indicating SAID and AK; after the BS verifies message 2, the BS replies with message 3, including AK, SAID, the old TEK (if any), and the new TEK, and HMAC. The old TEK includes the previously generated initialization vector, lifetime (in seconds), and sequence number of the data SA specified by SAID. The new TEK includes the next TEK's initialization vector, lifetime (in seconds), and the sequence number of the data SA specified by SAID, where the sequence number is incremented by one (modulo 4).

5.2.3 PKMv2

PKMv1 was first introduced in 2004 with the emergence of 802.16. After a number of security threats in PKMv1 were discovered, PKMv2 was introduced to fix them. The PKMv2 is composed of the authentication/authorization phase and the SA-TEK three-way handshake phase within the TEK transport/updates procedure. Before discussing the details of PKMv2, we first start with the comparison between PKMv1 and v2.

5.2.3.1 Comparison between PKMv1 and PKMv2

Comparing PKMv1 and PKMv2, the security capabilities of PKMv2 are significantly improved in many different ways.

- PKMv2 supports mutual authentication (i.e., both BS and MS authenticate each other) while PKMv1 protocol only allows unilateral authentication (i.e., BS authenticates MS, but not vice versa). By mutual authentication, the BS also has a certificate and can authenticate itself to the MS, which can remove the possibility of a rogue BS.
- Encryption algorithms in PMv2 are upgraded to more robust versions. PKMv2 supports a cipher suite stronger than DES-CBC (data encryption standard-cipher block chaining) in PKMv1 for protecting the MPDU origin's authenticity, integrity, and privacy. Note that regarding data security, PKMv1 provides no data authentication, no per packet integrity protection and no replay protection. PKMv2 supports message authentication based on CMAC, an alternative to the HMAC-based one in PKMv1, as well as encryption based on AES-CCM. A detailed list of the algorithms is shown in Table 5.3. However, some people argue that the security features are

Table 5.3. Comparison of PKMv1 and PKMv2.

	PKMv1	PKMv2
Authentication direction and method	One-way (BS authenticates MS) RSA based	Mutual RSA based/EAP based
Keys involved in authorization	AK	RSA based: Pre-PAK, AK EAP based: MSK, PMK, AK
TEK encryption	3DES, AES, RSA	3DES, AES, RSA, AES-ECB/KEY-WRAP
Data encryption	No encryption, DES-CBC, AES-CCM	No encryption, DES-CBC, 3DES, AES-CCM/CBC/CTR
MAC management frame integrity	No MAC, HMAC	No MAC, HMAC/CMAC
Security association type	Unicast	Unicast, GSA for multicast, MBSGSA for MBS
Others		Pre-authentication for handover

over strengthened in PKMv2, degrading the computing efficiency of the MS.

- PKMv1 suffers from replay attacks as there is no key liveness guarantee in the key exchange protocol and thus is vulnerable to the man-in-the-middle attack. PKMv2 includes a 64-bit random number in the authorization request and the reply messages to prevent these attacks.
- A single manufacturer credential based on X.509 certificate in PKMv1 is too limiting. On the contrary, PKMv2 supports multiple user credentials by using flexible protocols like EAP. PKMv2 allows using two well-known standards, RSA, and EAP, for authentication and authorization.
- In PKMv1, AK derivation is only BS's job and the MS does not have a chance to participate. But, in PKMv2, either preprimary AK (pre-PAK) or master session key (MSK) is sent first, from which AK can be derived at the MS and the BS, independently. These separate steps were included to protect the AK more securely.
- In PKMv1, no authorization SA is defined while data SA is explicitly defined. However, in PKMv2, both SAs are defined separately.

In addition to the security features listed above, PKMv2 was designed by considering additional security-related features and performance requirements in deployment, e.g., fast SA establishment, low latency key exchange, secure fast handover, and full backward compatibility with the existing solutions.

5.2.3.2 PKMv2 Operations

For the operation of PKMv2, many diverse authentication mechanisms can be employed. Specifically, RSA and EAP can be used in the authentication procedure in many different ways, such as RSA, RSA + EAP, EAP, and EAP-inEAP. Therefore, the AK derivation has become much more diverse and secure compared to PKMv1.

Key Hierarchy Since there are two authentication schemes in WiMAX, one based on RSA and the other based on EAP, keying materials are also managed in two ways. The key hierarchy in Figure 5.6 shows how keys are defined and generated in PKMv2.

There are three types of keys that form the roots of key hierarchy: the pre-PAK yielded by the RSA-based authorization process, the MSK yielded by the EAP-based authentication process, and the multicast and broadcast service authorization key (MBSAK) from which keys used to protect MBS

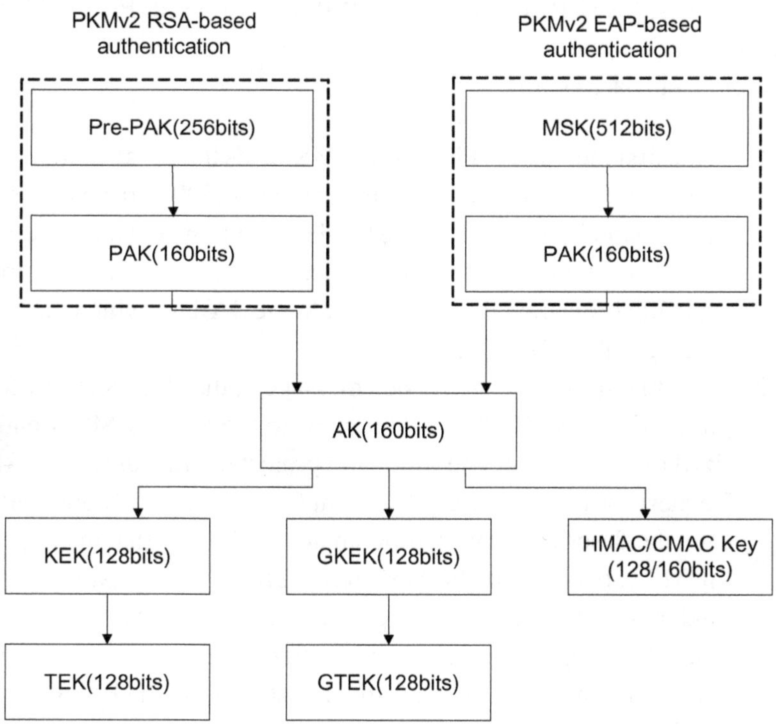

Fig. 5.6 Key hierarchies in PKMv2.

traffic are derived. Other keys, AK, KEK, group key encryption key (GKEK), TEK, group traffic encryption key (GTEK), MBS traffic key (MTK), message authentication keys for HMAC/CMAC, are derived from these three types of keys.

RSA-based authorization As in PKMv1, the RSA-based authorization in PKMv2 is also based on the public key infrastructure (PKI), which holds the standard X.509 certificate. For RSA-based authorization policy, PKMv2 RSA-Request, RSA-Reply, RSA-Reject, and RSA-Acknowledgment messages are used to share the pre-PAK. The BS sends the pre-PAK to the MS after encrypting it using the public key contained in the MS certificate. The pre-PAK is used to generate the PAK, which is in turn used to generate AK.

Once AK is securely generated at both MS and BS, both stations derive, independently, two keys, KEK and HMAC/CMAC key, from the AK. The HMAC/CMAC is used for authenticating the TEK request and reply messages. Similar to the RSA-based authorization in PKMv1, the MAC management messages, PKM-REQ, and PKM-RSP, are also used. Figure 5.7 shows the authentication and key exchange procedure based on RSA in PKMv2. Operation details are as follows:

1. In the first message in Phase 1, the MS sends the BS an Authentication Information message containing the X.509 certificate issued by the manufacturer itself. The BS determines whether the requesting MS is authorized for basic unicast services and/or additional statically provisioned services (i.e., Static SAIDs), which the MS's user has subscribed for.
2. After the message 1, MS sends message 2, the RSA Request for a pre-PAK and SAID. The SAID maps to the SA that MS is authorized to adopt during the following protected procedure. The RSA Request message includes the manufacturer-issued X.509 certificate, the description of cryptographic capabilities that the requesting MS supports, and the MS's basic CID of the MS, and a 64-bit random number. An MS's cryptographic capabilities are presented as a list of cryptographic suite identifiers. The basic CID, which is equal to the primary SAID, is the first static CID the BS assigns to an MS during initial ranging.

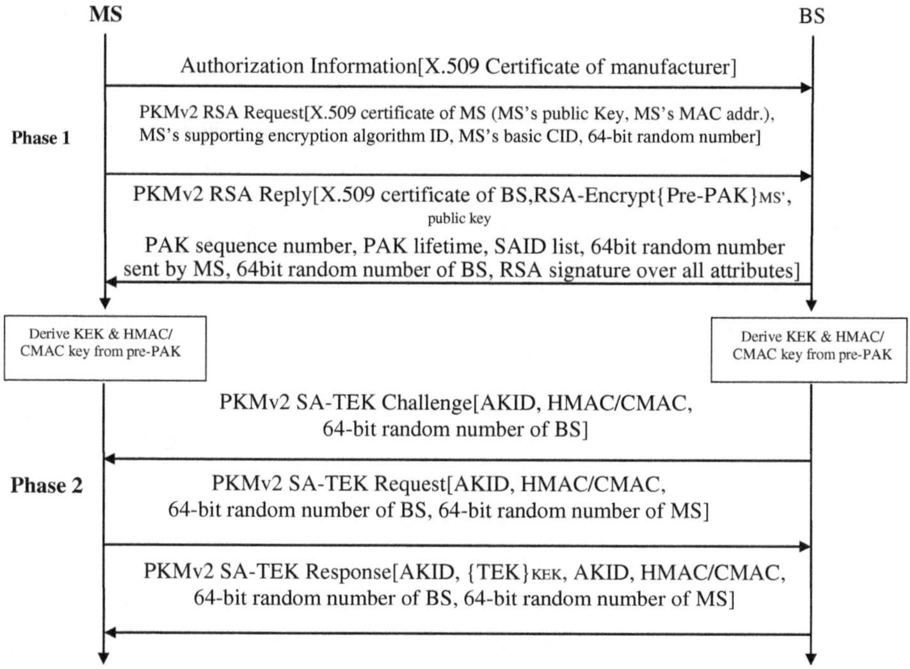

Fig. 5.7 Authentication and key exchange based on RSA in PKMv2.

3. In response to an RSA Request message, the BS performs the following actions. The BS validates the identity of the requesting MS's and determines the encryption algorithm and protocol support that it shares with the MS. If the MS is authenticated, the BS generates a pre-PAK for the MS, encrypts it with the public key of the MS, and sends it back to the MS in an RSA Reply message. The authorization reply message includes the BS certificate, the pre-PAK encrypted by the MS's public key, a 4-bit sequence number of pre-PAK, the lifetime of pre-PAK, and a list of SAIDs, a 64-bit random number of the MS, a 64-bit random number of the BS, and the RSA signature on the message. It also describes the properties of the single primary and zero or more static SAs by which the MS is authorized to obtain keying information. It is noted that the RSA Reply identifies the Primary SA and static SAs, but does not identify any dynamic SAs.

4. The MS and the BS derive PAK from the received pre-PAK, the AK from the PAK, and the KEK and HMAC/CMAC key from the AK, consecutively.
5. The BS starts the SA-TEK 3-way handshaking by sending the SA-TEK Challenge message to the MS to check if the MS possesses a valid AK.
6. The MS responds to this message by sending the SA-TEK Request message to the BS to request TEK.
7. The BS sends the SA-TEK Response message to the MS. Before the BS sends the encrypted TEK to the MS, it generates TEK and encrypts it using KEK.
8. The MS obtains TEK and uses it for encrypting/decrypting the data frames.

The AK is periodically refreshed by the MS by reissuing an authorization request to the BS. During reauthorization, however, the MS does not send authentication information messages. To avoid service interruptions during reauthorization, two AKs with overlapping lifetimes are maintained.

EAP-based authorization EAP is not a specific authentication mechanism, but a universal authentication framework defined by RFC 3748. It only defines message formats. Each protocol that uses EAP defines a way to encapsulate EAP messages within the messages of the protocol. EAP provides some common functions and negotiation of the desired authentication mechanism. Such mechanisms are called EAP methods, and currently about 40 different methods are in the list. Methods defined in IETF RFCs include EAP-MD5, EAP-OTP, EAP-GTC, EAP-TLS, EAP-IKEv2, EAP-SIM, and EAP-AKA. In addition, a number of vendor-specific methods and new proposals are also known. Among them, EAP-TLS is one of the widely deployed versions and is adopted as a basic mechanism for IEEE802.16e.

In the case of EAP-based authentication, both the AAA server and the MS first obtain a 512-bit MSK through the authentication process. The AAA server exchanges the MSK with the MS and the BS (or the authenticator), and then the MS and the BS derive independently the pairwise master key (PMK) from MSK, and then AK from PMK. The communication to the AAA server is done through the AAA carrier protocol, like RADIUS or Diameter protocols, most likely in a wired network.

If the MS (or BS) negotiates the authorization policy through the "authenticated EAP after EAP" mode (also called as the double EAP mode), it performs two rounds of EAP, and the authenticated EAP messages will carry the second EAP message after the first successful EAP-based authorization. The mode cryptographically binds the previous EAP authentication and the following EAP authentication session, while protecting the second EAP messages, which would enhance the protection capability. The same procedure is also applied when a RSA mutual authorization took place before the EAP exchange. In any of the cases, the EAP messages in the second EAP phase may be protected using 160-bit EIK derived from pre-PAK. Detailed authentication and key exchange procedures based on EAP in PKMv2 are shown in Figure 5.8 and explained in [1].

EAP-TLS In this section, we elaborate upon the EAP-TLS. In EAP-TLS, TLS is used to authenticate the TLS server to the client and the client to the TLS server. Figure 5.9 shows a detailed description of the second message "PKMv2 EAP-Transfer: EAP Conversation" in Figure 5.8. In the beginning of EAP authentication, the BS issues the EAP-Request/Identity message to the MS and the MS responds with EAP-Response/Identity, which contains the user's identity. The BS forwards the response message to the AAA server by encapsulating the EAP Start message within Diameter-EAP-Request.

Then, the AAA server issues a challenge message EAP-Request as a response to the BS and the BS sends a TLS-Start packet to the MS. The TLS Start message includes no data with the Start (S) bit set. Then, the TLS procedure starts when the MS sends a client_hello to the BS, which contains a random number, a sessionID, and a set of ciphersuites supported by the MS. The Session ID improves the efficiency in the case where a client repeatedly attempts to authenticate to the EAP server within a short period of time. Note that in TLS, a network is first authenticated to the MS, and then the network starts to authenticate the MS.

Next, the AAA server sends server_hello, its own certificate, request for MS' certificate, ServerKeyExchange, and Server_done message to the MS. The serverHello contains a random number, a sessionID, and a set of ciphersuites. The server_key_exchange is optional but must be included to allow the key exchange to take place where the certificate message contains a public key certificate chain for either a key exchange public key or signature public key.

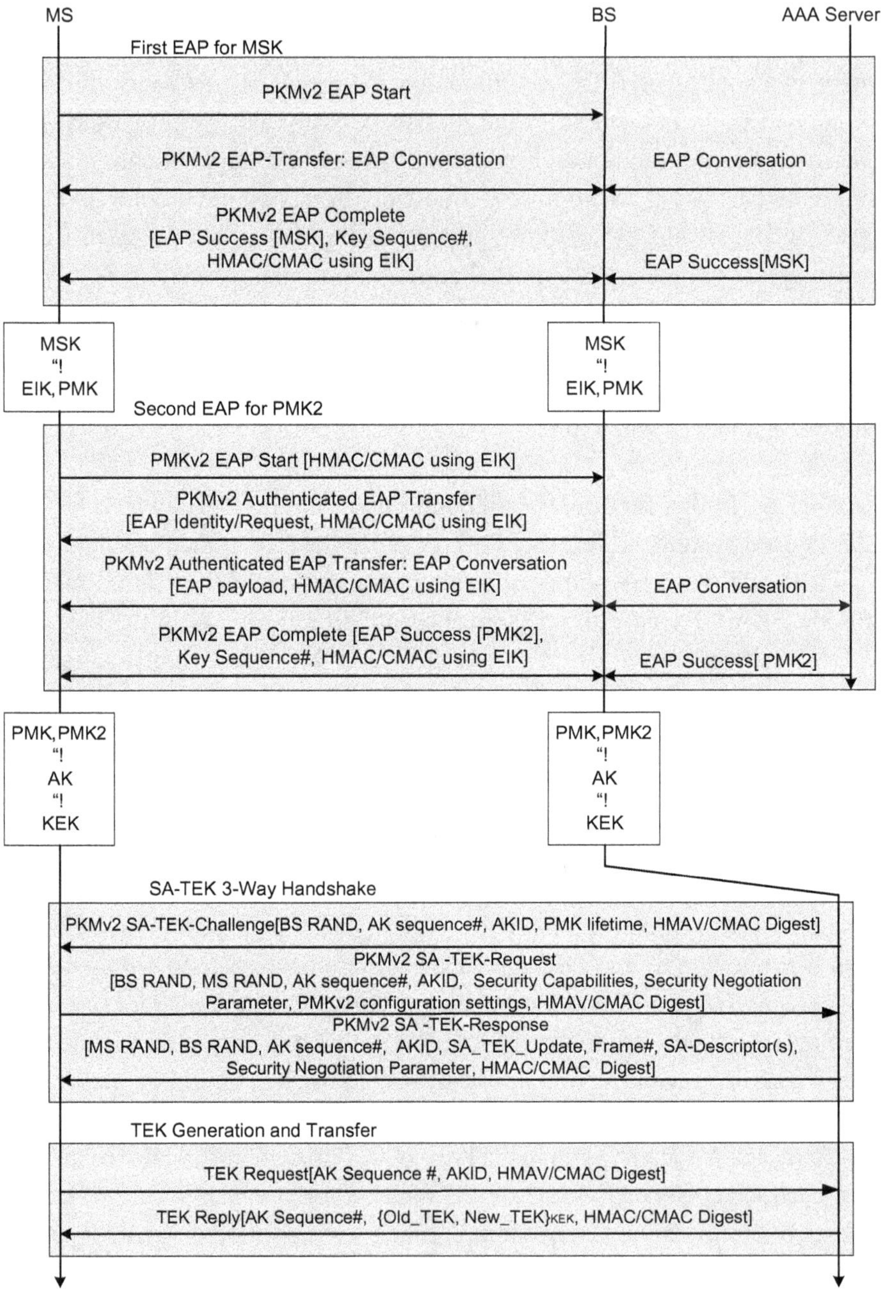

Fig. 5.8 Authentication and key exchange based on the double EAP mode in PKMv2.

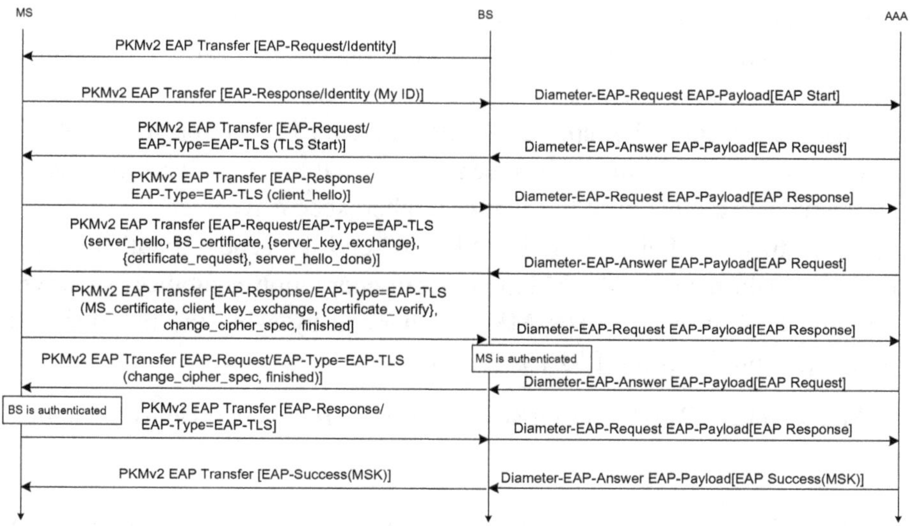

Fig. 5.9 EAP-TLS message flows between MS and AAA Server.

In the server_key_exchange, the Diffe-Hellman key exchange is involved. The certificate_request is also optional and is included when the server desires the clients to authenticate itself via a public key.

Then, the MS sends a message with client_key_exchange, its own certificate, its verification of AAA's certification notice, change_cipher_suite, and "finished" message encrypted with master key (MK). The MK is a key used within the EAP protocol and is created by the MS when it receives the server-Hello message. The MK is generated using its self-generated master secret, clientHello_random, and serverHello_random values. The MS sends the master secret to the BS after encrypting it by using the public key of the AAA server, which is contained in the BS_certificate. This encrypted information is included in the client_key_exchange message. At the same time, the hashed value of the MK is sent with the "finished" message. When the BS receives the master secret from the MS, it also generates the MK, and by using the MK, it confirms the hash value to verify the MS. Then, the BS also sends the hashed value of the MK along with the "finished" message.

The client_key_exchange completes the exchange of a shared master secret between the peer and the EAP server. The certificate_verify is included when the EAP server sent a certificate_request message in the preceding

EAP-Request packet and then the peer sends the certificate and certificate_verify messages. The certificate_verify contains the peer's signed authentication response to the EAP server. The change_cipher_spec is needed to change the current cipher spec and to negotiate the encryption specs with the AAA server. The current cipher spec is set when the TLS conversation begins and remains the same until the change_cipher_spec is received.

The attribute "finished" is included to send the sender's authentication response to the receiver. The MS indicates its encryption capability by encrypting "finished" with MK. The MK is generated by using the function TLS pseudorandom function (PRF) as follows:

$$\text{MK} = \text{TLS-PRF(PreMasterKey, "master secret" | clientHello.random| serverHello.random)}$$

where PreMasterKey is a 48-byte random number, which is generated by the MS and sent to the server by encrypting with server's public key that is made from the result of Diffe-Hellman key exchange and RSA-Encrypted Key.

Finally, the AAA server sends its responding change_cipher_spec and encrypted "finished" to MS. Then, the MS calculates the MSK using TLS-PRF as follows:

$$\text{MSK} = \text{TLS-PRF(MasterKey, "client EAP encryption" | clientHello.random| serverHello.random)}$$

Simultaneously, the AAA server gets MSK in the same way, passes it to the BS by protecting it in the RADIUS or Diameter protocol.

SA-TEK 3-Way Handshake As the second phase of PKMv2, the three-way handshake scheme is adopted to exchange the TEK between the MS and the BS. It is noted that before the three-way handshake begins, the BS and the MS both derive a shared KEK and HMAC/CMAC keys. It is also noted that the SA-TEK Challenge and SA-TEK Response messages from the BS to the MS are encapsulated in PKM-RSP management message, while the SA-TEK Request message is encapsulated in PKM-REQ message. The detailed process with possible attributes is described below.

1. During initial network entry or reauthorization, the BS sends the SA-TEK-Challenge message to the MS, which includes a random

number from the BS, KeySeqNo(AK sequence number), ID of the AK(AK identifier), key lifetime, and MAC for integrity. If the BS does not receive PKMv2 SA-TEK-Request from the MS within SAChallengeTimer, it will resend the previous message up to prespecified times. If retransmission reaches the prespecified limit, it will initiate another full authentication or drop the MS.

2. The MS sends PKMv2 SA-TEK-Request to the BS. The request message includes the MS_random number, BS_random number, which was received in the preceding message, KeySeqNo, ID of the AK, security capabilities included in the SBC-REQ message during the basic capabilities negotiation phase, SecNegParam (security negotiation parameters), PKM configuration settings, and MAC. Upon receiving this message, the BS can confirm that the supplied AKID refers to an AK that it has. The BS can verify the MS by comparing the random numbers. The BS must verify the MS's security capabilities encoded in the SecNegParam against the security capabilities provided by the MS through the SBC-REQ message. If the security parameters do not match, the BS reports the result to higher layers.
3. If the second message is verified, the BS will reply with PKMv2 SA-TEK-Response back to the MS. This message additionally includes SA-TEK-Update with TEK parameters, FrameNo indicating the frame number by which old PMKs and associated AKs should be discarded, and SADescriptors, which is valid only in initial entry.

Once the successful verification of the response has been done, the MS installs the received TEKs and associated parameters. In addition, the MS verifies the encoded BS's security negotiation parameters TLV against those provided by the BS through the SBC-RSP message. If they do not match, the MS reports the result to higher layers. In case the MS continues to communicate with the BS under this situation, the MS may adopt the security negotiation parameters encoded in the SA-TEK-Response message.

Key generation In authorization, the AK is the most essential keying material for further generation of other keys. The generation of AK in the

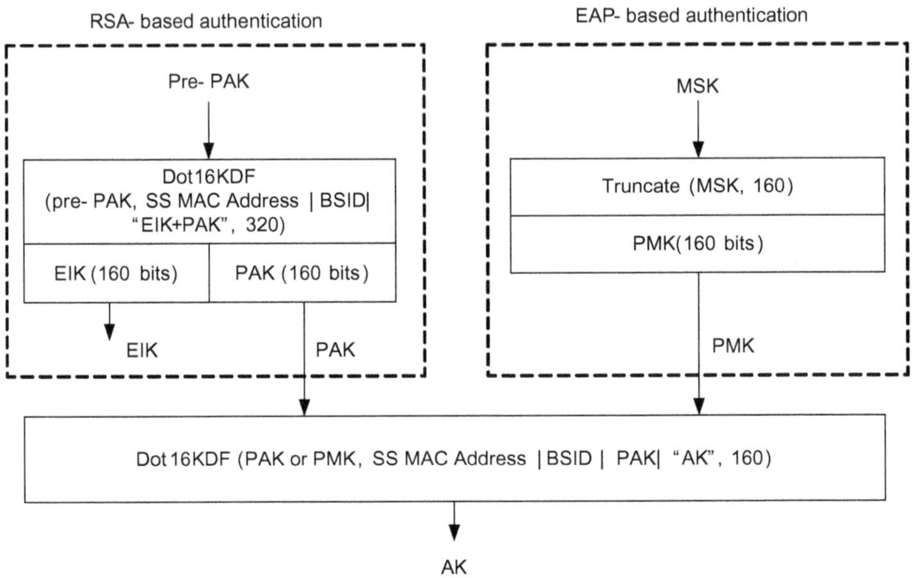

Fig. 5.10 Derivation of AK from Pre-PAK and MSK.[2]

RSA-based and the EAP-based authorization is summarized in Figure 5.10, where a|b denotes the concatenation of strings a and b, and Truncate(x,y) is a function that gives the rightmost y bits of a value x only if x is greater than or equal to y. In the RSA-based authorization case, EIK is also generated from pre-PAK for transmitting authenticated EAP payload.

Since RSA and EAP can be used in many different combinations in PKMv2, such as RSA + EAP and EAPinEAP besides the sole RSA and EAP, it is possible to have different ways of authentication and authorization. In other words, the AK can be derived in one of the three different ways depending on the authentication scheme used: (1) from PAK (2) from PAK and PMK, and (3) from PMK and PMK2. Figures 5.11 and 5.12 show the procedures of deriving the AK from RSA-based and EAP-based authorization, and EAP-based authorization and Authenticated EAP-based authorization, respectively. The notation ⊕ denotes the exclusive OR operation.

[2]IEEE Std 802.16e — IEEE Standard for the Software reviews ©2005 IEEE. All rights reserved.

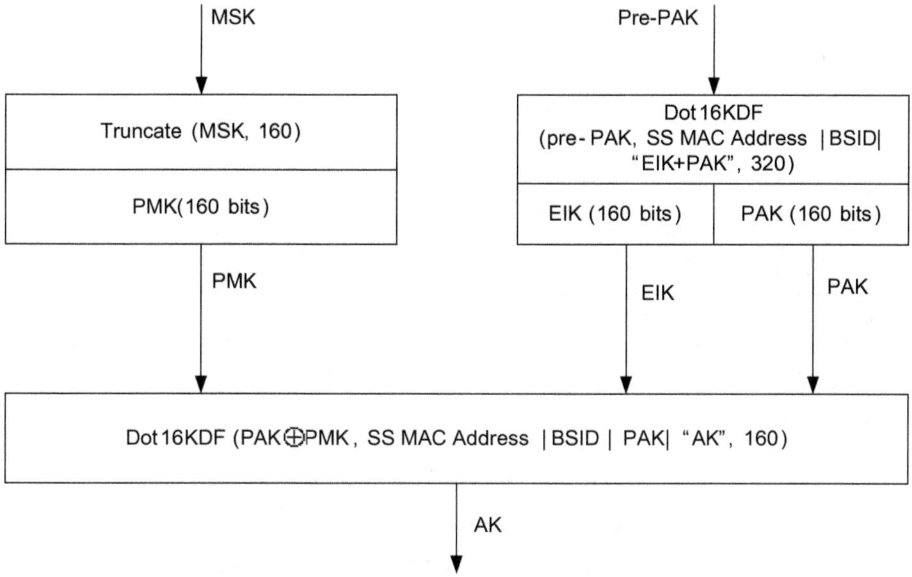

Fig. 5.11 Key derivation of AK from PAK and PMK (RSA-based and EAP-based authorization).

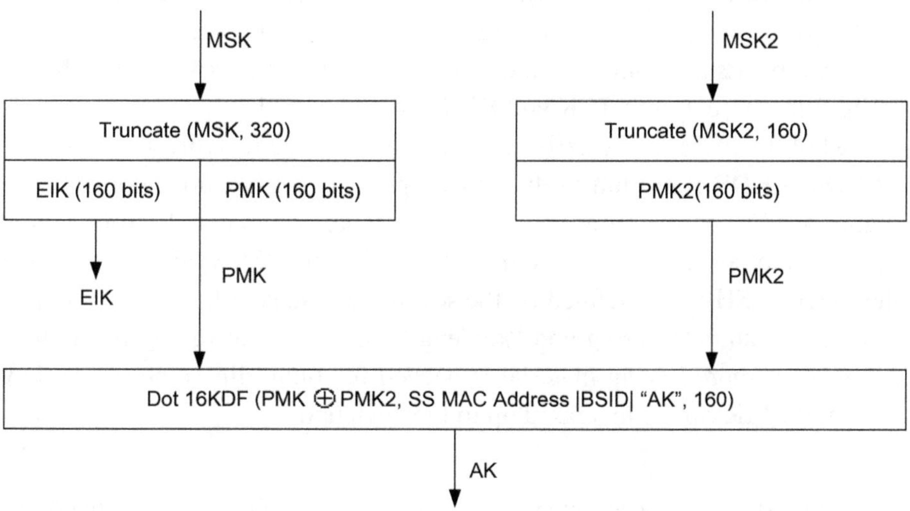

Fig. 5.12 Key derivation of AK from PMK and PMK2 (EAP-based authorization and authenticated EAP-based authorization).[3]

[3] IEEE Std 802.16e — IEEE Standard for the Software reviews ©2005 IEEE. All rights reserved.

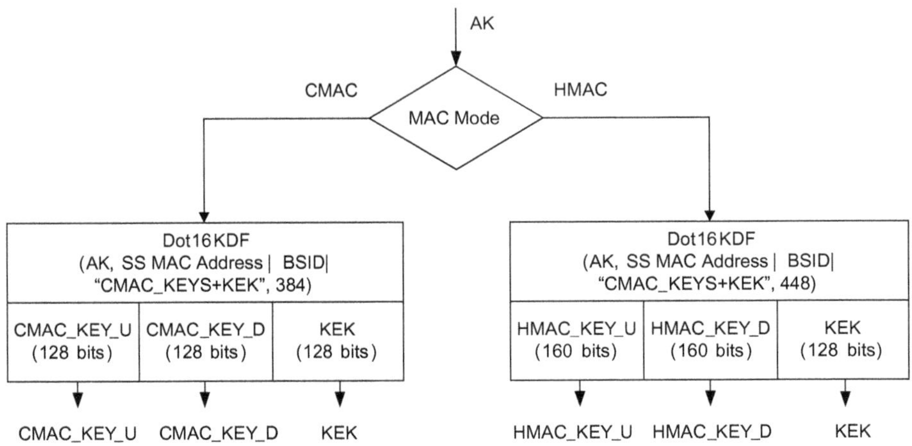

Fig. 5.13 Key derivation from AK.[4]

The HMAC/CMAC key and KEK derivation procedure from AK is shown in Figure 5.13, where HMAC/CMAC_key_U and HMAC/CMAC_key_D mean the HMAC/CMAC keys for uplink and downlink, respectively. TEK and KEK may be either 64 bits or 128 bits long. Security association employing any cipher suite with a basic block size of 128 bits uses 128-bit TEK and KEK. Otherwise, 64-bit TEK and KEK should be used.

All of the PKMv2 key derivations are based on the Dot16KDF algorithm. The Dot16KDF algorithm in the above figures is a CTR mode construction that is used to derive an arbitrary amount of keying materials from source keying materials. The flowchart in Figure 5.14 is the HMAC algorithm where the function SHA-1 is defined by the secure hash standard. In the algorithm, "astring" is an octet string and "keylength" is the length of key material to generate. If more keying material is needed for future link ciphers, the key length of PMK can be increased up to the keylength.

Key context The key context is represented by a set of parameters linked to a key in each hierarchy that defines the scope of a key. For example, key lifetime and counter values ensuring the same encryption will not be used more than once are the parameters in the set. When the context of the key expires, a

[4]IEEE Std 802.16e — IEEE Standard for the Software reviews ©2005 IEEE. All rights reserved.

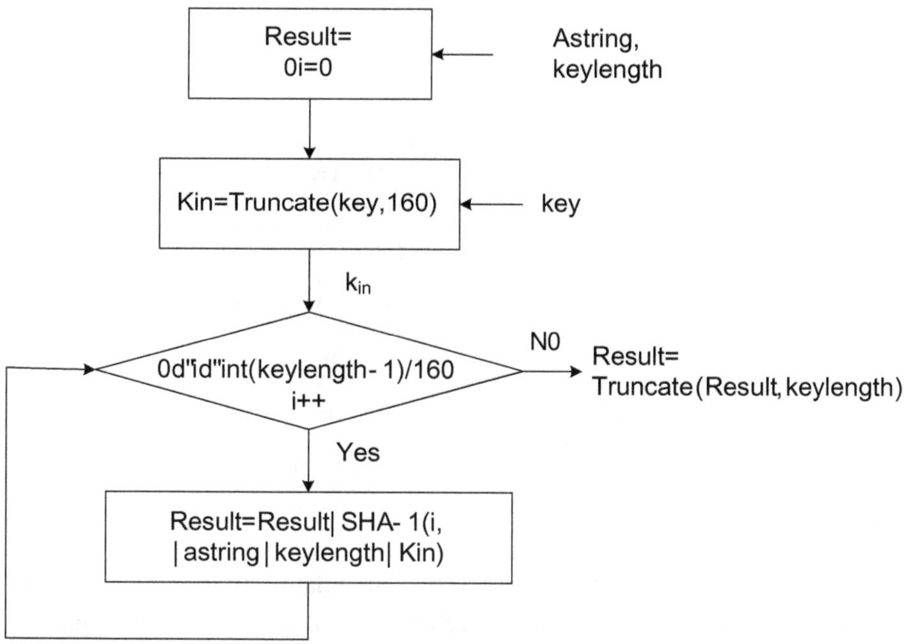

Fig. 5.14 Operation of Dot16KDF.

new key should be obtained. Table 5.4 summarizes the context belonging to a particular key.

Cryptographic methods

Data encryption The exchanged TEK is used to encrypt the MPDU. The data can be encrypted in four modes: (1) data encryption standard (DES) in cipher block chaining (CBC) mode; (2) advanced encryption standard (AES) in CTR with CBC-MAC (CCM) mode; (3) AES in counter (CTR) mode; and (4) AES in CBC mode. The encryption algorithm is selected by indicating the data encryption algorithm indicator in the cryptographic suite of an SA. In the generic MAC header, the MPDU is encrypted if the EC bit is set. In the IEEE802.16-2004 standard, two encryption modes, DES-CBC and AES-CCM, are included, while in the IEEE802.16e version, AES-CTR and AES-CBC modes were additionally included. AES-CTR (Counter) mode is mostly for MBS (multicast and broadcast service). Moreover, the AES Key Wrap algorithm with a 128-bit key was also included.

Table 5.4. Summary of key contexts (from [1]).

Contexts	Parameter
AK Context in PKMv2	AK
	AKID
	AK Sequence Number
	AK Lifetime
	PMK Sequence Number
	HMAC/CMAC_KEY_U
	HMAC/CMAC_PN_U
	HMAC/CMAC_KEY_D
	HMAC/CMAC_PN_D
	KEK
	EIK
PMK Context	PMK
	PMK sequence number
PAK Context	PAK
	PAK Lifetime
	PAK sequence number

DES is included in the CBC mode and is mandatory in IEEE802.16. In the DES-CBC mode, the initialization vector is calculated in the following way. In the downlink, CBC is initialized with XOR of the IV parameter included in the TEK keying material and the current frame number. In the uplink, the frame number of the frame where the relevant UL-MAP was transmitted is used instead of the current frame number.

AES-CCM is a block cipher that operates in the CTR mode, employing CMAC for data authentication and AES for data encryption. In the AES-CCM mode, an L-byte PDU payload is prepended with a 4-byte Packet Number (PN) and is appended with an 8-byte CMAC value. Thus, an L-byte data is augmented to as long as (L+12) bytes as shown in Figure 5.15. Note that the PN is not encrypted but is included in the message integrity check calculation. The PN value is set to one when an SA is established and when a new TEK is installed. After each PDU transmission, PN is increased by one.

TEK encryption TEK is protected by KEK. The BS encrypts TEK before transmitting it to MS in the Key Reply message. The BS encrypts the TEK-128 in the Key Reply messages it sends to the client MS when the TEK encryption algorithm identifier in an SA is equal to 0x04. For SAs using a ciphersuite employing 64-bit DES-CBC, the TEK in the Key Reply is encrypted by triple DES (3DES), using a two-key 3DES KEK derived from AK. For SAs using

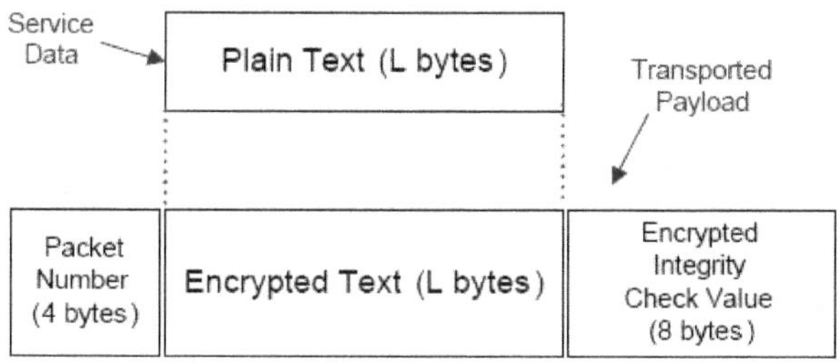

Fig. 5.15 MSDU payload format.[5]

a ciphersuite employing AES-CCM mode with a 128-bit key, TEK is AES encrypted using a 128-bit key derived from AK. The encryption is done by using the AES Key Wrap Algorithm, which is described in detail in the following figure. The algorithm accepts both a ciphertext and an integrity check value for encryption and returns a plaintext and the integrity check value after decryption.

TEK update Due to the lifetime of keys, the keys should be updated periodically to guarantee continuous freshness. Especially, TEK must be updated periodically for minimizing the possibility of keys being leaked to adversaries. The BS always maintains two sets of keying material per SAID, and sends both of them in its Key Reply messages. The keying material includes two KEKs and their remaining lifetimes. In order for the smooth transition of keys in case the lifetime of keys expires, two keys may be simultaneously active with overlapping lifetime in each SA. The lifetimes of the two keys overlap such that each key becomes active halfway through the life of its previous one and expires halfway through the life of the next one.

Utilizing the remaining lifetime of a currently active key, the MS can estimate when the BS will invalidate a particular TEK. This means the MS can schedule when to send the next Key Request messages so that the MS receives new keying material before the BS invalidate the current keying material in the MS. Furthermore in the AES-CCM mode, beside the Key Request scheduling algorithm, another update ensuring mechanism is employed, i.e., when more than half the available PN numbers in the 31-bit PN number space are exhausted, the MS schedules the next Key Request message.

[5]IEEE Std 802.16e — IEEE Standard for the Software reviews ©2005 IEEE. All rights reserved.

The following is the message flowchart of the TEK update process. As mentioned, the BS and the MS maintain two TEKs at all times through the exchange of Key Request and Reply messages. For an AK, two TEKs correspond to the older and the newer generations of keying material, respectively. The BS uses the two key differently depending on whether the keys are used for uplink and down link communication. The rules that the BS uses are as follows [14]:

1. The BS uses the older of the two active TEKs for encrypting downlink traffic.
2. The BS decrypts the uplink traffic using either the older or newer TEK, depending on which of the two keys the MS uses at the time.

Figure 5.16 illustrates the TEK update procedure between BS and MS. The BS manages two TEKs and uses the newer TEK when the older TEK expires. It is the responsibility of the MS that it periodically refreshes the TEK by requesting the BS to send a newer TEK. The MS sends a Key Request message to the BS to request TEKs of the SAID before the expiration of the newer TEK the MS currently holds and after the scheduled expiration of the older of the two TEKs. The time interval between the instant of sending the Key Request message and the end of the lifetime of the newer TEK is called *TEK grace time*. Thus, the Key Reply message with a newer key should arrive before the TEK grace time is over. The shaded portion of a TEK's lifetime describes the time interval when the TEK is specifically used to encrypt the traffic data.

This key management mechanism in the PKMv2 makes sense in the WiMAX environment, where the BS is assumed to have more computing capability and higher power than the MS. Note that, however, there are high communication overheads aroused by the key transport and some researches investigated how to reduce the high burden computation. Ender Yuksel claims that either of random numbers in the SA-3 way handshake can be omitted without degrading the security strength by utilizing the Lysa model analysis. For details, refer to [13].

5.3 Preauthentication and Handover Security

Mobility is one of the important features of the Mobile WiMAX system. Usually, continuous service is achieved by supporting handover from one cell

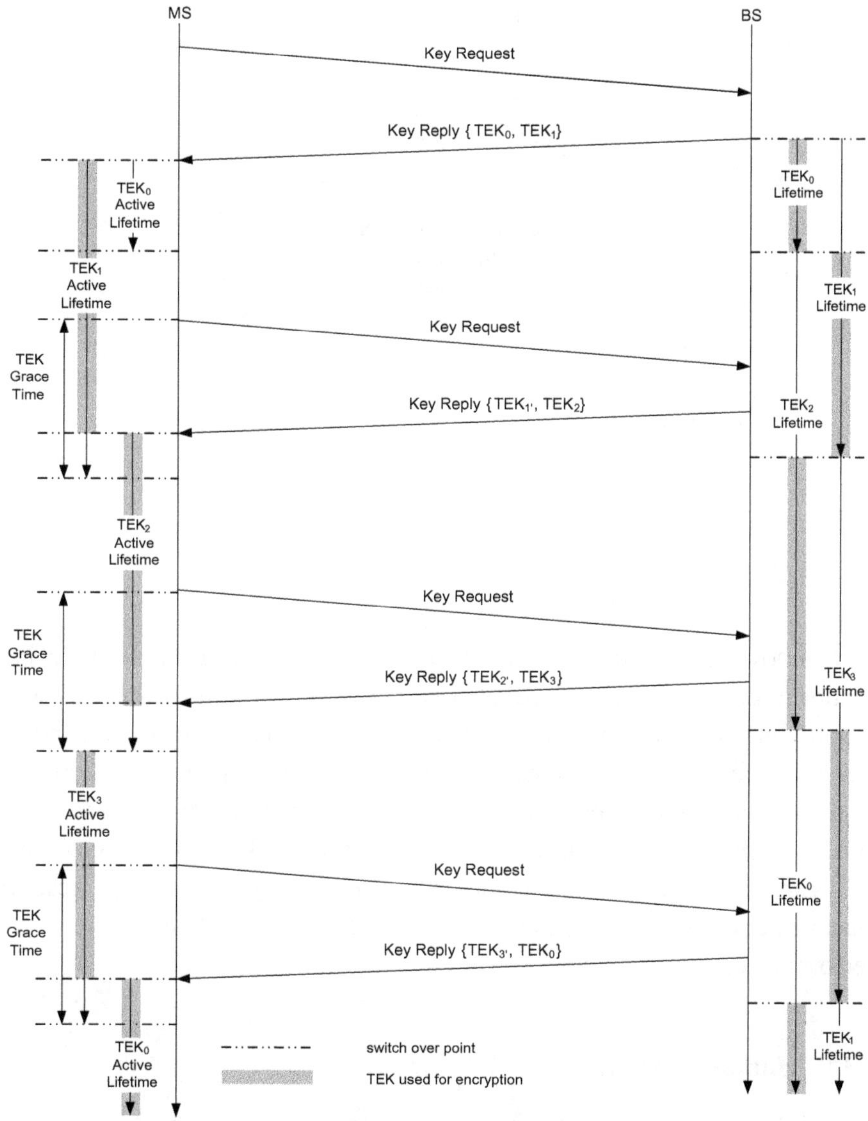

Fig. 5.16 TEK update procedure (from [1]).

to another. Handover is the process of changing either frequency, time slot or spreading code, or combination of them associated with the current connection during the call progress. It is often initiated either by crossing a cell boundary or by deterioration in quality of the signal in the current channel.

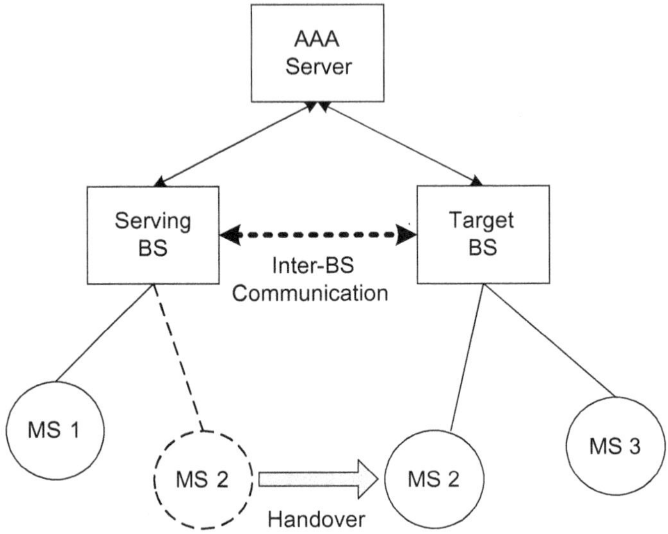

Fig. 5.17 Handover model.

Handover is divided into two broad categories: hard and soft. In hard handover, current resources are released before new resources are used while in soft handover, both of two sets of resources are used during the handover process. The reason why handovers are critical in cellular communication systems is that whenever handover occurs, negotiation for new parameters must take place among the MS, the current serving BS, and the new target BS as in Figure 5.17. Since handover includes many factors, e.g., resource management and security, the handover strategy can influence the overall performance.

5.3.1 Handover Model

IEEE 802.16e also supports seamless handover to allow the MS to switch from one base station to another at vehicular speeds without interrupting the connection. Handover schemes are required to guarantee the latency to ensure the real-time applications such as VoIP to perform without service degradation. There are three handover methods supported within the 802.16e amendment: Hard Handover (HHO), Fast Base Station Switching (FBSS), and Macro Diversity Handover (MDHO). While the HHO is mandatory, FBSS and

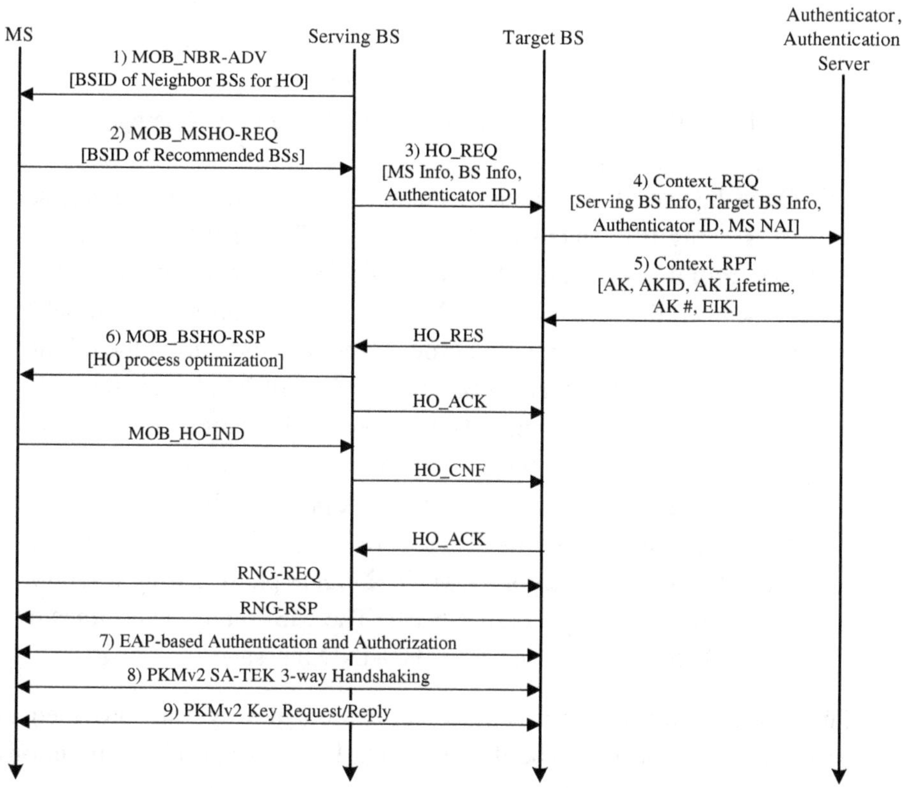

Fig. 5.18 Handover procedure in WiMAX.

MDHO are two optional modes. Figure 5.18 describes the overall handover procedures of WiMAX.

1. The MS receives the MOB_NBR-ADV message containing the information of neighbor BSs from the serving BS periodically.
2. When the MS decides to handover, it sends an MOB_MSHO-REQ message to the currently serving BS so that the BS dispatches a HO-Request message to the new possible target BSs. To select possible BSs, the MS scans neighbors and collect signal information, i.e., mean values of received signal strength index (RSSI), CINR, or delay. The number of BSs can be more than one.
3. The serving network sends Handover Request (HO_REQ) to target BSs.

4. The target BSs that received HO_REQ ask security context to authenticator or authentication server by sending Context_REQ.
5. The authenticator or authentication server may reply with AK, AKID, AK Lifetime, AK sequence number, and EIK using a Context_RPT message.
6. Once the two BS's exchange the context and verification messages, the serving BS can ensure that the target BS is ready to serve the MS from that time on. After receiving the handover acknowledgment from the target BS, the MS initiates the initial ranging and authentication/authorization procedures, as it did during the initial network entry. The serving BS replies MS with MOB_BSHO-RSP containing the indication of HO process optimizations. Especially, Flag Bit #1of HO process optimization indicates the omission of PKM Authentication phase except TEK phase during current reentry processing (Steps 7 and 8 in Figure 5.18). Moreover, Flag Bit #2 indicates the omission of PKM TEK create phase during re-entry processing (Step 9 in Figure 5.18). The MS finally determines a target Bs by sending the MOB_HO-IND message.

The flag bits in the HO process optimization message are provided to support seamless handover. The flag has eight kinds of optimization options for reducing the delay in network reentry process. As mentioned, among the flag bits, Flag Bit #1 and #2 are related to WiMAX security. It must be noted, however, that even if these two bits allow fast authentication by reducing the handover latency for real-time service, it is possible to bring up critical security vulnerabilities. Thus, HO optimization is the result of the trade-off between handover performance and communication security, and further research is required to improve both of the goals simultaneously.

5.3.2 Preauthentication

From the security point of view, in order to achieve the seamless handover, it is important to maintain the security context during the handover. In the meantime, from the performance point of view, it is also important to finish the handover within a time-bound schedule so that the QoS of the traffic can be maintained. For both of these two goals, more flexible and fast key management scheme that omits the EAP/RSA authentication process may need to be devised in order to spare the time consumed during the keying establishment in the usual PKM process.

As one possible approach, preauthentication has been considered to support the fast MS reentry authentication to the neighboring cell without long overhead. By anticipating the target BS, an MS may rely on preauthentication for an accelerated reentry at a particular BS. The selection of potential target networks is necessary for transferring security context before handover, which is inherently based on the probabilistic prediction. Preauthentication enables the MS and target BS to establish a new AK. However, preauthentication was not finally included in the standard due to some controversial issues. Although the specific mechanism for preauthentication is out-of-scope of the IEEE802.16e specification, there have been made a few proposals to enhance the performance.

For example, the draft specification P802.16e/D5a defines a preauthentication procedure and message formats. Preauthentication is based on the principle that a centralized AAA server establishes a shared private key MK between itself and the MS using an EAP method, and populates multiple BS's with a pairwise master key (PMK) that is derived from the MK and the identities of the BS and MS (e.g., MAC address). Thus the MS is able to compute the PMK for any BS, since it is based on the MK it possesses, its own identity, and the publicly known identity of any target BS it may hand over to. This foreknowledge of the PMK enables a fast handover, making formal key exchanges unnecessary.

Although preauthentication was studied actively during the standardization period, this part was not included in the final IEEE802.16e specification. One possible reason was that the experts believed they can possibly achieve the goal of preauthentication with other handover primitives. That is, they supposed that the function of preauthentication can be supported by the exchange of Context_Request/Receipt messages containing security context and HO process optimization without defining additional management frames and procedures.

5.4 WiMAX Network Architecture and Security Issues

WiMAX Forum has published a series of documents that describe the WiMAX network architecture reference model, reference points, and protocols and procedures. The documents serve as an architecture framework for WiMAX deployments and ensure interoperability among various WiMAX equipments and operators. Based on the IEEE802.16 PHY and MAC specifications, the architecture framework and network reference model (NRM) defined by the WiMAX forum accommodate diverse usage models for different end-to-end aspects. Thus, the architecture refers to appropriate IETF RFCs and IEEE Ethernet standards besides IEEE 802.16 standard and its amendments.

5.4.1 Entity Definitions and Network Reference Model

The NRM is a logical representation of the network architecture, consisting of the logical entities, e.g., ASN, CSN, ASN-GW, NSP, etc. The following definitions are from [17]:

- Access service network (ASN): a complete set of network functions needed to provide radio access to a WiMAX subscriber. ASN comprises network elements such as one or more BS(s), and one or more ASN Gateway(s). An ASN may be shared by more than one connectivity service networks (CSN). For authentication, ASN may also contain an authenticator or a network access server (NAS).
- ASN-GW (ASN gateway): Typically a layer-2 traffic aggregation point within an ASN. Additional functions that may be part of the ASN gateway include intra-ASN location management and paging, radio resource management and admission control, caching of subscriber profiles, and encryption keys, AAA client functionality, establishment and management of mobility tunnel with base stations, QoS and policy enforcement, foreign agent functionality for mobile IP, and routing to the selected CSN.
- Connectivity service network (CSN): A set of network functions that provide IP connectivity services to the WiMAX subscriber(s). CSN may comprise network elements such as routers, AAA proxy/servers, user databases, Interworking gateway MSs.
- Network access provider (NAP): A business entity that provides WiMAX radio access infrastructure to one or more WiMAX network service providers (NSPs). A NAP implements this infrastructure using one or more ASNs.
- Home network service provider (H-NSP): An operator or business entity that has service level agreements (SLA) with WiMAX

subscribers, authenticates and authorizes subscriber sessions (in-network or roaming scenarios) and services the subscriber's account (charging and billing). To support roaming services, an H-NSP may have roaming relationships with other NSPs.
- Visited network service provider (V-NSP): A visited operator or business entity that a roaming subscriber uses for access to WiMAX services. A visited NSP may have roaming relationship with subscriber's home NSP. The visited NSP provides AAA traffic routing to home NSP. Depending on WiMAX services requested and the roaming agreement between home NSP and visited NSP, the visited NSP may provide some/all WiMAX services to roaming WiMAX subscriber or provide data/control traffic routing to home NSP.

The ASN provides the following mandatory functions [17]:

- Layer 2 connectivity with the MS
- Transfer of AAA messages to WiMAX subscriber's H-NSP for authentication, authorization and accounting
- Network discovery and selection of the subscriber's preferred NSP
- Replay functionality for establishing layer 3 connectivity with an MS (i.e., IP address allocation)
- Radio resource management

On the other hand, some of the important functions that a CSN can provide are as follows [17]:

- MS IP address and endpoint parameter allocation for user sessions
- Internet access
- AAA proxy or server
- Policy and admission control based on user subscription profiles
- ASN-CSN tunneling support
- Billing and inter-operator settlement.

Each of the above entities represents a grouping of functional entities that may be implemented in a single physical entity or can be distributed over multiple functional entities. While how to organize and distribute the functions within the ASN is manufacturer-dependent and an implementation choice, the network working group in WiMAX Forum has defined three ASN interoperability profiles — Profiles A, B, C [18]. Note that the network architecture framework permits decoupling of access architecture from IP connectivity services. This means that the IEEE 802.16 standard does not conceive the network elements of the CSN.

Over the functional entities, the NRM identifies reference points (RPs) over which interoperability is achieved between functional entities. A reference point, denoted by R1-R8, is a conceptual point between two groups of functions that reside in different functional entities on either side of it. These functions include various protocols associated with an RP. Figure 5.19 shows a network reference model and several interoperability reference points. The brief description for each reference point is as follows [17]:

- R1 consists of the protocols and procedures between MS and ASN as per the air interface (PHY and MAC) specifications.
- R2 consists of protocols and procedures between MS and CSN associated with authentication, services authorization, IP host configuration management, and mobility management.
- R3 consists of the set of control plane protocols between ASN and CSN to support AAA, policy enforcement, and mobility management capabilities.
- R4 consists of the set of control and bearer plane protocols originating/terminating in various functional entities of an ASN that

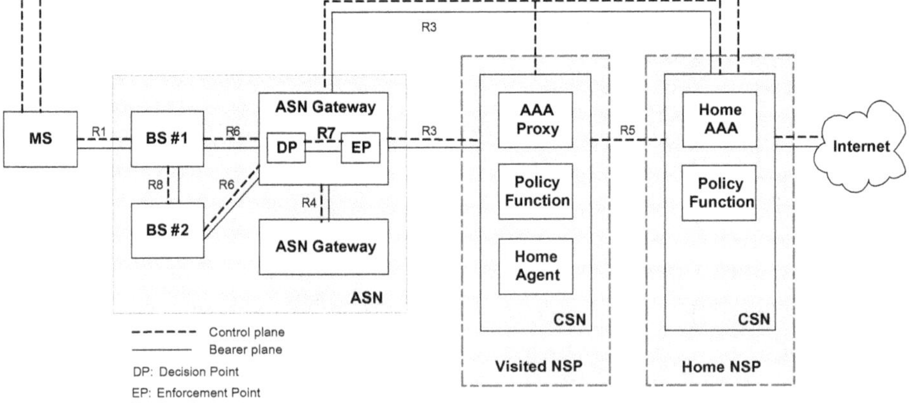

Fig. 5.19 WiMAX network reference model [17].

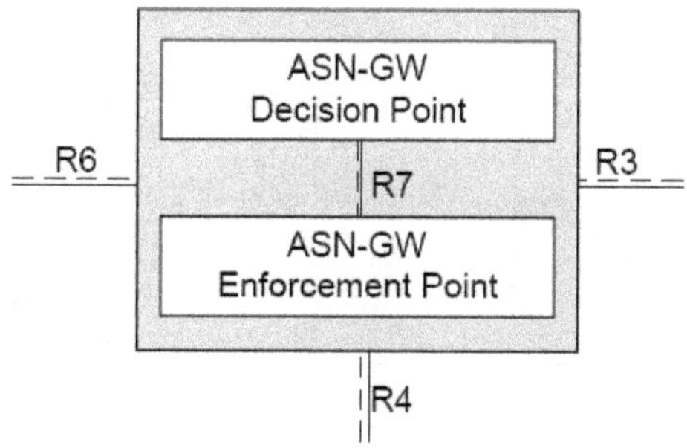

Fig. 5.20 Reference point R7 within ASN-GW [17].

coordinate MS mobility between ASNs and ASN-GWs. R4 is the only interoperable RP between similar or heterogeneous ASNs.

- R5 consists of the set of control plane and bearer plane protocols for internetworking between the CSN operated by the home NSP and that operated by a visited NSP.
- R6 consists of the set of control and Bearer Plane protocols for communication between BS and ASN-GW.
- R7 consists of the optional set of Control Plane protocols as shown in Figure 5.20, e.g., for AAA and Policy coordination in the ASN gateway, as well as other protocols for coordination between the two groups of functions identified in R6. R7 optionally decomposes the aggregated ASN-GW, which is a logical entity that represents an aggregation of control plane functional entities, into the decision point and the enforcement point.
- R8 consists of the set of control plane message flows and optionally bearer plane data flows between BSs to ensure fast and seamless handover.

In Figure 5.19, the ASN gateway is connected with V-NSP and H-NSP. H-NSP contains an actual Home AAA server while V-NSP contains AAA proxy. Note that in this case, the AAA framework is compatible with the AAA 3-party scheme — an MS as a "supplicant," an "authenticator" in ASN, and an "authentication server" in the AAA backend.

5.4.2 Security of RPs in NRM

In order to ensure end-to-end security in the NRM, security must be considered at each RP. The security must be ensured at both the lower layer (PHY,

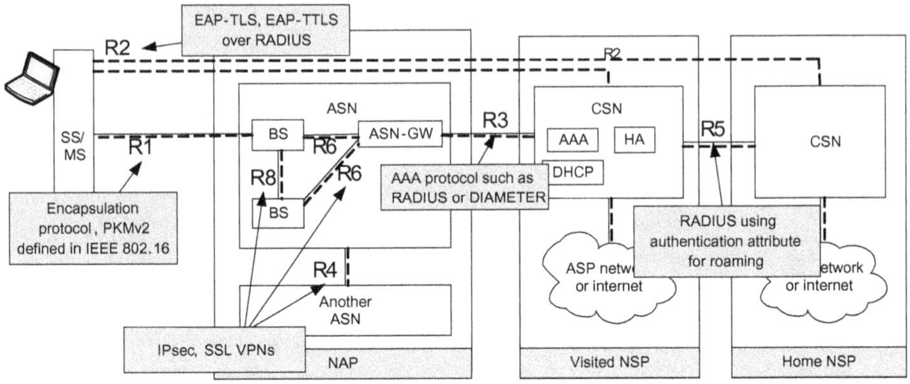

Fig. 5.21 Example deployment of security mechanisms in WiMAX.

MAC, or network layer) and the higher layer, because the security at these two groups of layers is complementary. The layered security must be deployed at each RP. Figure 5.21 shows an example deployment of security mechanisms at each reference point, which can be used as a guideline to WiMAX deployments.[6] In the following, some general guidelines for the deployment are described [17]:

- R1: It is the reference point which IEEE 802.16 primarily concerns. The 802.16 management connection over R1 is authenticated, integrity and replay protected in the MAC layer upon successful device authentication. All the subsequent R1 messaging over these connections can rely on this lower-layer cryptographic security. On the other hand, transport connections may not be crypto-protected at all. For that, any signaling protocol and data traffic that run on these connections shall provide their own security when necessary.
- R2: It is assumed that the lower-layers are insecure and the signaling protocols and data traffic shall provide their own security when necessary. Generally, EAP methods, such as EAP-TLS or EAP-TTLS, are used for device and network authentication.

[6]Among the reference points, R7 is not defined in the specifications since this reference point is an optional internal connection in an ANS-GW.

- R3: This reference point may not have an end-to-end secure channel. It can be used in such a case when a mobile user undergoes handover under the Mobile IPv4 with authentication extensions.
- R4: This reference point is for supporting inter-ASN mobility. It has an end-to-end secure channel, including privacy. The channel security may be implemented using physical security, IPSec or SSL VPNs, etc. The VPN end points may be collocated with the R4 end points, or be on-path between the two to ensure end-to-end security.
- R5: This reference point supports roaming across multiple different NSPs. It is a practical assumption that the lower-layers are insecure since two NSPs are different. It can be made secure by adopting, for example, RADIUS authentication attribute.
- R6: This connection consists of intra-ASN bearer paths and IP tunnels for mobility events. This reference point has an end-to-end secure channel, including privacy. The channel security may be implemented using physical security, IPSec or SSL VPNs, etc.
- R8: This reference point has an end-to-end secure channel, including privacy. The channel security may be implemented using physical security, IPSec, or SSL VPNs, etc.

The WiMAX Forum strongly recommends protecting the RPs and interfaces between all interconnected RADIUS client, proxy, and server entities. Thus, any specific protection method must be determined by deploying parties. Moreover, a variety of different data stores used by RADIUS, e.g., user's identity store, policy store, and an accounting information store, must be securely maintained. But the procedures for provisioning, maintaining, and securing these stores are also left out-of-the-scope of the specification.

5.4.3 EAP-based Authentication

From the viewpoint of WiMAX Forum, PKMv1 provides support only for device authentication while PKMv2 provides a flexible solution for both device and user authentication between MS and home CSN. In order to ensure the end-to-end aspect of authentication and authorization, the forum requires the IEEE 802.16e EAP-based PKMv2 to be used between MS and ASN. Within ASN, Intra-ASN security describes additional steps to transfer EAP messages

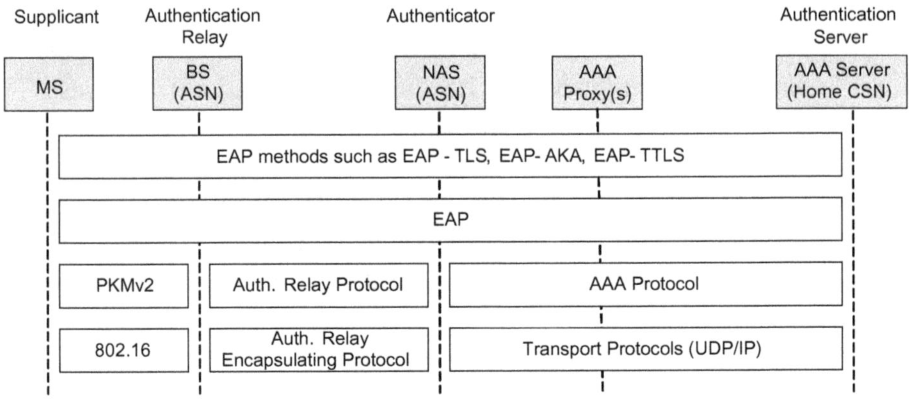

Fig. 5.22 PKMv2 user authentication protocols [18].

and keys among the entities. Between the AAA server and the authenticator in ASN, EAP runs over RADIUS.

5.4.3.1 User Authentication

In order to authenticate a user (see Figure 5.22), EAP transfer is made over the IEEE 802.16 air interface between MS and BS in ASN. EAP messages may be forwarded by the BS to an authenticator over the authentication relay protocol, depending on the location of authenticator in the ASN. The EAP messages are encapsulated within AAA protocol packets by the AAA client on the authenticator and forwarded via one or more AAA proxies to the AAA server in the CSN of the home NSP where the supplicant is subscribed to.

5.4.3.2 Device Authentication

WiMAX Forum also requires that the EAP must be used for device authentication. EAP methods used for device authentication are required to be able to generate the MSK and EMSK key. Regarding the time when the device authentication will be performed, the operator must decide it as a policy that the MS must learn during the MS provisioning process. That is, the policy may dictate not performing device authentication at all, performing it after power-on, and so on. A typical policy is to perform both device and user authentication at each power-on only. Thus, until the next power-on, user-only authentication may be sufficient to authorize IP access of the MS.

5.4.3.3 User and Device Authentication

In case both user and device authentication need to be performed, it can be done in the following two ways: using EAP tunneling or combined authentication.

1. A tunneling EAP method (e.g., EAP-TTLS) performs two EAP authentications over what appears like a single EAP session. There are inner and outer methods, where the inner EAP method can be used for user authentication and the outer one can be used for device authentication.
2. Combined authentication is a joint authentication of a device and a user as one single EAP when both authentications are based on a preshared key (PSK) and terminate at the home CSN. Combined authentication generates a combined identity, with which one PSK-based EAP authentication is performed. This optional optimization aims at reducing the authentication setup latency.

5.4.3.4 Detailed Authentication Procedure

The procedure based on PKMv2 during the initial network entry of the MS is shown in Figure 5.23. This figure is basically the same as Figure 5.8, but it is redrawn based on the NRM in Figure 5.19. Additional explanation for the first three steps is as follows:

1. During the SBC negotiation, MS and ASN negotiate the PKM version, PKMv2 capabilities and authorization policy including requirements and support for device authentication. After the successful 802.16 link setup, the Authenticator begins the EAP sequence.
2. Depending on the authenticator location, i.e., BS or ASN-Gateway, the message may be transferred over the authentication replay protocol across the R6 RP.
3. The MSK is established at the MS and the Home AAA server. By this step, the authentication part of the authorization flow (and the involvement of the generic EAP layer) is completed.
4. The rest of the steps are the same as described in Figure 5.8.

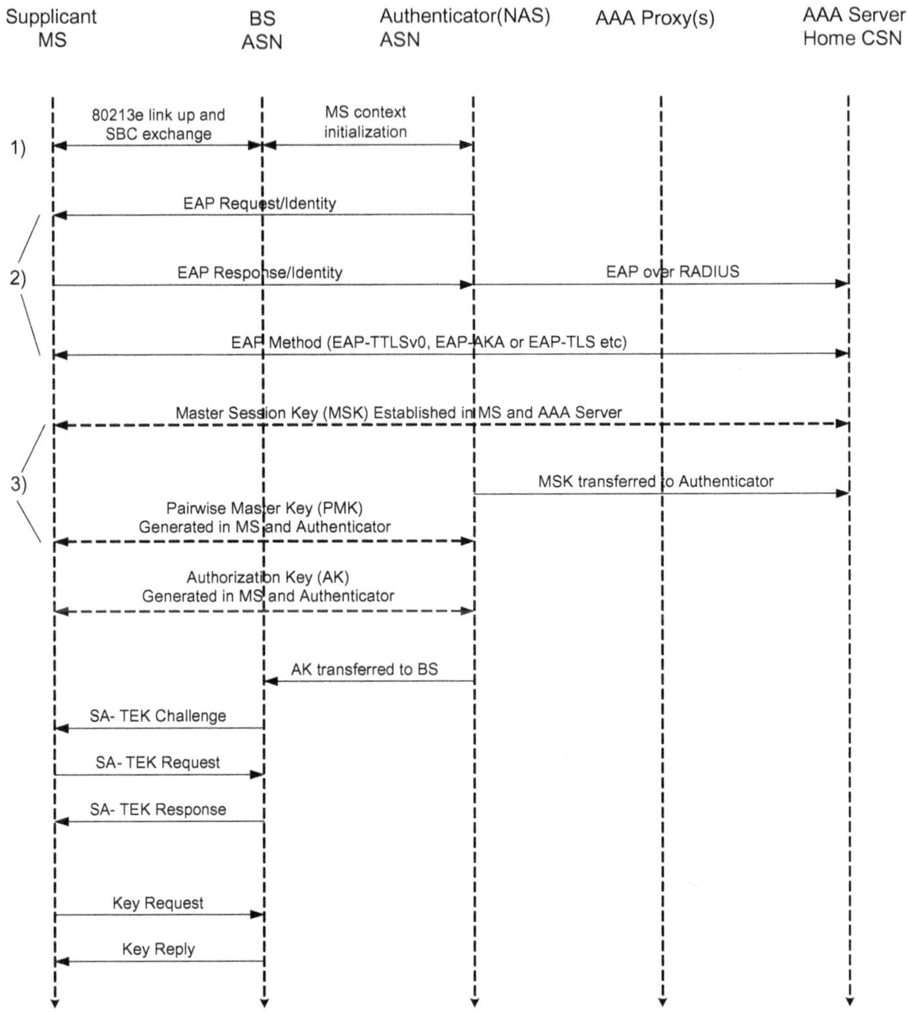

Fig. 5.23 Authentication model in WiMAX Forum [17].

5.4.3.5 ASN Security Architecture

Since an ASN may have network entities such as a BS and an ASN gateway, its inner architecture can be further classified into authenticator, authentication relay, key distributor, and key receiver depending on their functions. The authenticator is an entity that authenticates a user or a device, and the authentication relay is what simply relays EAP packets without message snooping

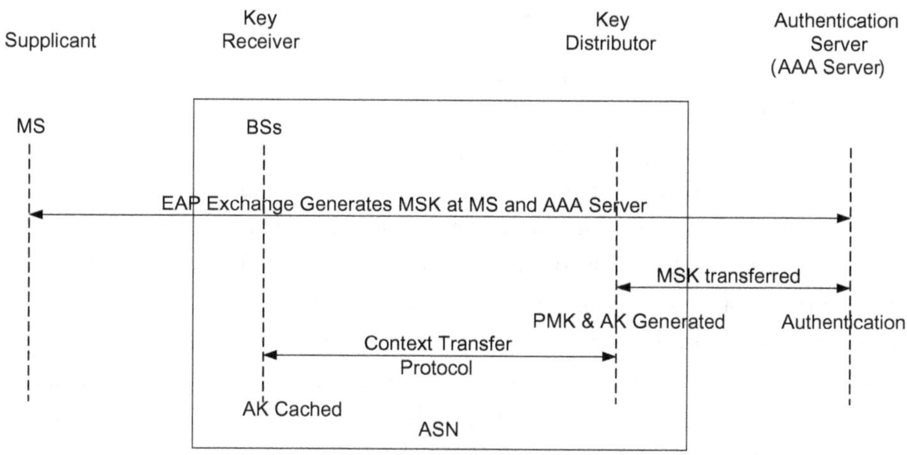

Fig. 5.24 AK Transfer inside the ASN [17].

or modification via an authentication relay protocol. The key distributor is the functional entity that holds the MSK and the PMK obtained from EAP exchange. It also derives AK and creates AKID for an ⟨MS, BS⟩ pair and distributes AK and its context to the key receiver in a BS via context transfer protocol. The key receiver is the key holder for AK and generates other keys specified by IEEE802.16e from AK.

Figure 5.24 shows a key transfer relationship among functional entities. It is noted that the context transfer protocol is responsible for securely transferring the keying material, i.e., valid AKs and their contexts, from the key distributor to the key receiver in the target BS in case the MS hands over. The protocol is a two-message exchange consisting of an optional Request message and a mandatory Transfer message. The key receiver sends the Context_REQ message for requesting a new AK from the key distributor. For this message, the key distributor sends the Context_RPT message or indicates a failure. The identity of the key distributor is determined based on the MS identifier in the Context_REQ message. Upon receiving the Context_RPT message, the BS encrypts the AK, AK Lifetime, AK Sequence Number, CMAC_KEY_COUNT, and EIK, etc., and stores them locally for future use. In document [17], a few sample scenarios are presented on how the Context_REQ and Context_RPT exchange is triggered. One possible way is by issuing the MOB_HO-IND, RNG-REQ, or MOB_MSHO-REQ message.

5.4.3.6 Additional Comments on Security in NRM

From the aspect of NRM, IEEE802.16e specifications only cover the security between MS and BS, i.e., RP R1 in Figure 5.19. It means that for other RPs, e.g., R3 and R6, no specific countermeasures are specified. Thus, it is up to the developer on how to provide security for those domains.

As a countermeasure for this security problem, the authors in [12] propose a key exchange method based on the PKI. Using the PKI, the communicating parties obtain the public key and its certificate of their communicating correspondents. In order to do this, they assume that the WiMAX devices are certified by public authority, and can verify the certificates and a certificate chain. If a communicating party (e.g., BS) would like to exchange messages with its correspondent (e.g., ASN-GW), the BS first generates a session key (e.g., ASN-TEK and ASN-CSN-TEK) and send the key and the authority's certificate to the ASN-GW after encrypting the key using the correspondent's public key. When the ASN-GW receives the message, it first tries to verify the certificate and check the time validity using the timestamp. This scheme can be applied in a similar way to other RPs and thus, can establish secure connection in WiMAX NRM.

5.5 Threats and Solutions

In wireless communication networks, a number of threats can exist in the two lower layers, PHY and MAC, due to its broadcast nature. Data can be easily intercepted, and communication can be intervened by both signal interference in the physical layer and MAC message eavesdropping and modification in the MAC layer. A number of research results have revealed many types of threats and vulnerabilities residing in the WiMAX system standard. For example, Barbeau describes an extensive analysis on the security threats of WiMax/IEEE802.16 with quantitative models [11]. The author analyzed the threats with respect to their likelihood of occurrence, their possible impact on individual users and system, and the global risk they represent.

Even if the newest version of WiMAX, IEEE802.16e-2005, fixed some of the fatal security holes found in the previous version, several new security holes have been discovered after the advent of amendment. In Refs. [12, 13], the authors analyzed the security aspects of PKM and revealed the PKM scheme to be insecure even after the amendment. Thus, there is still much room for research in this field as new requirements and newly uncovered threats will continuously emerge in a constantly varying environment.

In this section, we review some possible threats and vulnerabilities in the IEEE802.16 standards known in the open literature.

5.5.1 Physical Layer Threats

In the physical layer of WiMAX, two subframes, the uplink and downlink subframes, are used for transmission. In the uplink subframe, the time division multiple access (TDMA) scheme is adopted, where each of the time slots is designated for individual MS served by the BS. The time division duplex (TDD) downlink subframe consists of two fields, i.e., the control information field and the data field. The control field includes preamble and maps: Preamble contains frame synchronization information, while the maps announce the start point of the transmission characteristics of the following data bursts. Due to the broadcast nature and TDMA scheme, the communication at the physical layer is vulnerable to two kinds of attacks, jamming and scrambling.

5.5.1.1 Jamming

The jamming attack is achieved by introducing a source of noise, which is strong enough to surpass the original signal, reducing the channel capacity and affecting on the signal quality. Although the jamming attack can be malicious or unintentional, it is not difficult to apply the attack. The countermeasure is also relatively simple and it is not difficult to resist the jamming attack by augmenting the original signal (i.e., using a transceiver with higher power), increasing bandwidth, using spreading spectrum and so forth.

5.5.1.2 Scrambling

The scrambling attack is a sort of the jamming attack, but is for shorter intervals of time and targeted to specific frames or parts of frames. Usually, the scrambler selectively aims at control or management information with the purposes of affecting the normal operation of the network. The problem caused by scrambling is serious mainly to the time-sensitive messages, such as channel measurement report requests or responses, because the attack may force the messages to be retransmitted as a result. It is relatively more difficult for attackers to apply scrambling than jamming, because he/she needs to interpret control messages and send noise during specific intervals. Detection of scrambling is also tough because the intermittent nature of the attack may seem to be normal noises. The scrambling attack has been studied for WiFi/802.11 systems, but remains as an open problem in the case of WiMAX/802.16.

5.5.2 MAC Layer Threats

According to the IEEE802.16e security sublayer, the payloads in the MAC layer are encrypted, while the MAC header that facilitates the operation of the MAC layer is transmitted in the clear text format. Thus, although confidentiality and integrity of the payload are guaranteed by the encryption methods, the management messages are not encrypted and vulnerable to eavesdropping attacks. (Note that authentication of management messages is provided for integrity checking as an option.) Additionally, there is a potential for denial of service (DoS) attacks because a long procedure of authentication operations may be vulnerable to the flooding attack with a high number of messages to authenticate. The impact could be quite high to a user with lower computational resources available for processing authentication requests.

5.5.2.1 Vulnerabilities in PKMv1

Besides the vulnerabilities described above, the original IEEE802.16-2004 has been known to have many other kinds of vulnerabilities. They were mainly related to the Ranging Response vulnerability, Auth Request Invalid vulnerability, and rogue BS vulnerability [12, 15]. In the first two cases, the problem is originated from the absence of nonintegrity checking capability within control messages, and it was possible for an attacker to change the contents of the message as he/she liked to. In the rogue BS vulnerability case, the rogue BS pretends to be a legitimate BS, confusing a set of the MS's trying to get services from it. This attack was possible due to the unilateral MS authentication, but this vulnerability has been removed by adopting the mutual authentication between MS and BS in the newer version of the standard.

More specifically, in the case of the Auth Invalid vulnerability, an attacker may exploit the unauthenticated control messages during the exchange of Key Request and Reply message. Specifically, Auth Invalid message is generated either internally by MS when the Key Reply or Key Reject messages are not properly authenticated or externally by BS when the BS receives an invalid Key Request message from the MS. In either of these two cases when MS receives an Auth Invalid message, it will change its state from "Authorized" to "Reauthorization Wait". If the wait timer expires before MS receives new messages from BS, MS sends a Reauth Request message to BS for obtaining new key information. However, during the wait time period, if MS receives an Auth Reject message, it will move to the "Permanent Authorization Failure" mode and be pushed into a silent state. In this state, MS finishes all the connections and become ready to respond to any command from BS. Since the Auth Invalid message is not protected by the message authentication codes, is not identified by an identifier, and only carries the error codes with short description on the failure causes, it is easy for the attacker to modify the message contents intentionally. However, with the amendment in IEEE802.16e-2005, this problem has been solved and all PKM messages are now protected using message authentication with HMAC/CMAC.

5.5.2.2 Vulnerabilities in PKMv2

While many of the advanced features have been included to strengthen the security of WiMAX, it is still vulnerable to many potential attacks. The authors in [12] describe some of the vulnerabilities, i.e., disclosure of security context during network entry, lack of secure communication in an access network, and vulnerability during handover. As can be seen, some of the threats have not been solved with effective counter measurements and further researches are needed to find the solutions.

Describing the security during network entry, every MS is required to perform the initial ranging, SBC negotiation, PKM authentication/authorization, and registration procedures during the initial network entry stage. Although the initial entry stage is one of the essential parts for establishing a connection between MS and BS, many of the parameters and contexts are not protected and sent in a clear text format. Specifically, the SBC negotiation parameters and PKM security contexts are not secured confidential. While the standard provides message authentication and encryption schemes, they are only applicable to data traffic after the initial network entry, not to the control messages during the initial network entry process. Thus, it is always possible for a malicious user to eavesdrop or change the contents as he/se wishes to do during the initial network entry process, for which a countermeasure should be devised.

As a possible countermeasure to protect the exchanged control messages, the authors in [12] propose to use the Diffie-Hellman (DH) key agreement scheme. Noting that the DH key agreement scheme ensures to share an encryption key based on the prime number "p" and its primitive root "q". When MS receives a UL-MAP message including ranging codes during the initial ranging procedure, the MS selects one of them and sends it back to BS. At this time, the BS acknowledges that the ranging code has been accepted by sending an RNG-RSP message. At this stage, it is also possible to use the code for generating a prime number "p" for the DH key agreement scheme. At the same time, the MS also generates the other variable "q" and a public/private key pair, and send them to the BS. When the BS receives a public key of the MS and primitive root, it verifies them and sends its public key to the MS, too. In this manner, the MS and the BS should be able to share the DH variables and public key with each other, and can then generate a key, pre-TEK, for securing the exchanged control messages afterwards during the initial network entry procedure.

5.6 Conclusions

In this chapter, we reviewed the security features of the IEEE802.16e, a mobile WiMAX standard. We described the general architecture and operation of the WiMAX security systems and functions in terms of encryption, authentication, and key management. In more detail, we focused on PKM, PKMv2, and EAP-based authentication.

Since its first emergence in 2005, Mobile WiMAX has been evolving to IEEE802.16e Rev.2 in 2009 and to IEEE802.16m in 2010–2011, which is under consideration for future standard IMT-Advanced. Correspondingly, WiMAX Forum released the Mobile WiMAX network specification R2.0 in 2010. 802.16m may include more diverse features such as location based service, support for femto-cells, self-organization, multicarrier support, and enhanced multicast and broadcast services, besides many advanced features in PHY and MAC layers.

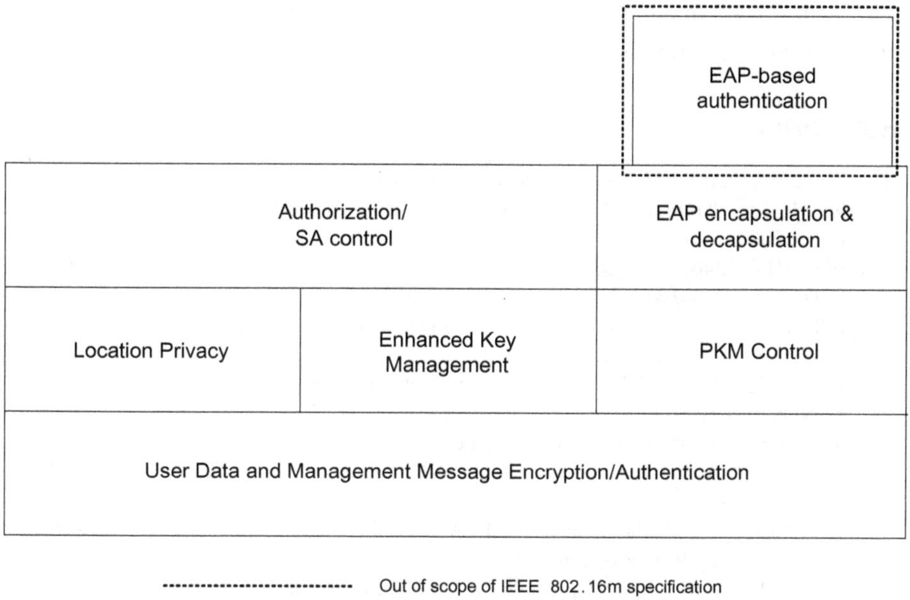

Fig. 5.25 Generic security architecture of IEEE802.16m [19].

The generic security architecture of IEEE802.16m is shown in Figure 5.25. The architecture is expected to be divided into two logical entities: security management entity and encryption entity [19]. Security management entity includes the functions such as overall security management and control, EAP encapsulation/decapsulation, privacy key management ver.3 (PKMv3), which defines how to control security components such as keys, authentication and SA control, and location privacy. On the other hand, encryption and integrity protection entity includes functions such as transport data encryption/authentication processing, control message authentication processing, and control message confidentiality protection.

Since the standardization of 802.16m is still on-going, the security part will possibly be updated until the last stage. Due to the additional features of 802.16m, the detailed security architecture will become more complicated compared to previous versions. Especially, by the introduction of location-based service and femto-cell environment, there will be more complex security and privacy issues. It is expected that the IEEE802.16m security architecture will be designed in a more robust and reliable manner to counteract the future possible security attacks.

References

[1] IEEE Standard for Local and Metropolitan Area Networks. *Air Interface for Fixed and mobile Broadband Wireless Access Systems*. IEEE Std 802.16e [S]. New York: IEEE Press, 2006.
[2] IETF RFC 2246, The TLS Protocol Version 1.0.
[3] IETF RFC 3748, Extensible Authentication Protocol (EAP).
[4] IETF RFC 4187, Extensible Authentication Protocol Method for 3rd Generation Authentication and Key Agreement (EAP-AKA).
[5] IETF RFC 2026, EAP Tunneled TLS Authentication Protocol (EAP-TTLS).
[6] K. Nyberg, Cryptographic Algorithms for UMTS, ECCOMAS, 2004.
[7] MS-CHAP v2, http://technet.microsoft.com.
[8] K.-T. Do, TEK update procedure during HO and reentry, IEEE 802.16 Broadband Wireless Access Working Group Proposal, May 2007.
[9] D. Johnston, Pre Authentication for PKMv2, IEEE 802.16 Broadband Wireless Access Working Group Proposal, June 2004.
[10] M. Leech, Security Requirements for WiMAX, Presentation slides.
[11] M. Barbeau, WiMAX/802.16 Threat Analysis, ACM Wireless Networks, 2005.
[12] T. Shon and W. Choi, *An Analysis of Mobile WiMAX Security:Vulnerabilities and Solutions*. Springer, 0302-9743.
[13] E. Yuksel, Analysis of the PKMv2 Protocol in IEEE 802.16e-2005 Using Static Analysis, 2007 IMM-THESIS-2007-16.
[14] L. Nuaymi, *WiMAX: Technology for Broadband Wireless Access*. Wiley, 2007.
[15] D. Johnson and J. Walker, Overview of IEEE 802.16 security. *Security & Privacy Magazine*, 2(3), 40–48, 2004.
[16] WiMax Forum Network Architecture (Stage 2: Architecture Tenets, Reference model and Reference Points) [Part 0].
[17] WiMax Forum Network Architecture (Stage 2: Architecture Tenets, Reference model and Reference Points) [Part 1].
[18] WiMax Forum Network Architecture (Stage 2: Architecture Tenets, Reference model and Reference Points) [Part 2].
[19] IEEE802.16 IMT-Advanced Evaluation Group Coordination Meeting, Overview of IEEEP802.16m technology and candidate RIT for IMT-advanced, January 2010.

6

Security for Other Systems

There are several other systems besides EPS and WiMAX. In this chapter we cover security for few other topics of importance. These are (i) machine-to-machine (M2M) communications, (ii) home (evolved) NodeB (H(e)NB), the Femto solution from 3GPP, and (iii) multimedia broadcast and multicast services (MBMS).

6.1 M2M Communications and Security

6.1.1 Introduction to Machine-to-Machine Communication

Machine-to-machine (M2M) communication is a form of data communication that involves one or more entities that do not necessarily need human interaction. M2M uses a device to capture an event, which is relayed through a network to an application, which translates the captured event into meaningful information. This is accomplished through the use of telemetry, the language machines use when in communication with each other. Such communication was originally accomplished by having a remote network of machines relay information back to a central hub for analysis, which would then be rerouted into a system like a personal computer.

However, modern M2M communication has expanded beyond a one-to-one connection and changed into a system of networks that transmits data to personal appliances. The expansion of wireless networks across the world has made it far easier for M2M communication to take place and has lessened the amount of power and time necessary for information to be communicated between machines. These networks also allow an array of new business opportunities and connections between consumers and producers in terms of the products being sold.

6.1.2 Use Cases

(1) Traffic cameras

Traffic cameras with cellular connectivity may be installed in locations such as motorway overpasses or remote stretches of roadway. Cameras may also require simultaneous secure local WLAN connectivity to the next camera down the road, for example, when measuring average speed. It will be necessary to securely provision these cameras with subscription credentials. When cameras are deployed over a large area, it may be necessary to be able to select a carrier for a given camera after it has been deployed, and this selection process must be properly secured. Secure postdeployment changes in subscription data, including change of operator, will also be needed.

(2) Remote metering

A change of utility by the residential customer may also require a change in operator. The utility itself may switch operators, requiring a change to many meters dispersed over a large geographical area in a limited timeframe. The management of these changes may require complex accounting mechanisms. Without the ability to remotely manage subscription credentials, a service person may need to visit each affected device. For commercial applications, obtaining physical access to deployed devices may be expensive, because of geography, extreme environmental conditions, or the need to interrupt a manufacturing process (e.g., petrochemical refining). Therefore, remote means of managing subscription data could be needed.

(3) Vending machines

Vending machines dispersed in many locations are subject to regular attacks on their contents, which increase the threat to other items of value in the machine. Vending machine connectivity may come from a Home NodeB or 3GPP I-WLAN access within the customer premises. A change in the customer's choice of network operator may require an update to subscription data in many vending machines in a short time. The vending machine supplier may also change its preferred network operator.

(4) Asset/cargo tracking

Asset and cargo tracking will often require that the M2M equipment be placed in areas where physical access is difficult. Such placements would be part of a service provider's attempt to resist theft and tampering with the M2M equipment. These placements can make it difficult and costly to physically manage subscription credentials in the M2M equipment. As noted in TR22.868, this is "practically impossible" under the current solution, and a means of securely re-provisioning MCIM applications over the air would therefore be very beneficial.

6.1.3 Security Issues on M2M Communications

Because the devices involved in M2M are dispersed in diverse environments, they are fundamentally vulnerable to both physical and cyber attacks. Some of the security issues that can be raised in M2M communications are as follows:

(1) How to prevent theft of and tampering with subscription credentials?

(2) How to initially provision a new M2M equipment with a new USIM application from an operator of M2M subscriber choice?

- How can the M2M subscriber select his chosen home operator after the M2M equipment has been delivered from the supplier?
- How can the M2M equipment be remotely and securely provisioned with a new MCIM application from the operator chosen by the M2M subscriber?
- How can the home operator ensure the trustworthiness of the M2M equipment?

(3) How to change subscription to a different operator?

6.1.4 Security Threats on M2M Communications

(1) Physical attack

Physical attack includes theft, vandalism, and insertion of valid authentication tokens. Beside human-to-human communications, the communication devices are used to be left alone for some months or years because most of the M2M communication is operated without human's interaction. If there is no management for the machine, however, the machine device is vulnerable for physical attack. The physical attack not only causes the financial loss, but also may lead to the secret leak. Moreover, the physical demolition and secret theft without the machine administrator's awareness may cause the more severe problems.

(2) Compromise of credential

The machine may be compromised due to the weak authentication algorithm, physical intrusion, malicious cloning of authentication tokens, or side-channel attacks. The target of most attacks of this type is machine communication identity module (MCIM).

(3) Configuration attack

Configuration attack includes fraudulent software update/configuration changes, misconfiguration by the owner, subscriber, or user, or the compromised access control lists.

(4) Protocol attack on the device

This attack impacts the device directly. The protocol attack includes the man-in-the-middle attack on first network access, denial-of-service attack, and compromising a device exploiting weaknesses of active network services.

(5) Attack on the core network

This attack is the main threat to the mobile network operator: Impersonation of devices, traffic tunneling between impersonated devices, misconfiguration of the firewall in the modem/router/gateways, and DoS attacks against the core networks. They may also include changing the device's authorized physical location in an unauthorized fashion or attacks on the radio access network, using a rogue device.

(6) User data and identity privacy attack

This attack includes eavesdropping for other user's or device's data, masquerading as other user/subscriber's device, user's network ID, or other confidential data revealed to unauthorized third parties.

6.1.5 Security Requirements for M2M Communications

(1) Data confidentiality

In M2M communications, the transmitted data may include sensitive information, i.e., source location, utility bill, USIM information, etc. If the sensitive data are not encrypted, the data can be revealed to the adversary by man-in-the-middle attack or eavesdropping. Moreover, due to the lack of the human's interaction, the operator may not aware that the sensitive data leak. To prevent the problems, transmitted data should be encrypted. Furthermore, for the efficient encryption, M2M communication may employ a group key.

(2) Data integrity

The transmitted data may include the metering information (for example, pressure of pipeline, temperature of the house, etc). Users receive these metering data via networks. If the measured data are modified illegally or deleted, the user may suffer damage. For example, when the machine that measures the temperature of the house is compromised by the malicious code, the fire occurs in the house, and the machine reports the false temperature information to a subscriber, the subscriber must suffer severe damage. To prevent the problem with data integrity, the transmitted data should be proven that it is not modified. For this, message authentication with signature is needed.

(3) Device integrity

Machines are vulnerable to the physical attacks, i.e., theft and vandalism. The network cannot prevent the problems, but can detect them in order to deactivate the machine equipments' service and the related USIM. In addition, often theft or vandalism vulnerable machines are stationary after the first initial installation and activation. The fixed machines can be utilized to improve the detection of the theft or vandalism. If a known machines move or are destroyed, it can be concluded that the machines have been stolen or vandalized and thus the account deactivated.

(4) System availability

If some machines are attacked by the DoS attack, the availability and productivity of the machines becomes worst, so that the accessibility for system resource and information is lowered. If the machine device is unavailable when the subscribers have to use, the subscribers may receive a serious damage. Hence, the protection mechanisms for availability are required.

(5) Privacy

Because the M2M devices are closely related to the human's life, the transmitted data may include the sensitive data, i.e., user's identity number, password, or location. If the data are revealed to the illegal third party, the privacy can no longer be guaranteed. Hence, the security mechanisms that prevent the privacy attack are required.

(6) Trace prevention

For the mobile device, the location information or routing information of the device or the device's owner can be revealed. To prevent it, the security mechanisms that provide both mobility and prohibition from tracing are required. For this purpose, the schemes that can hide the source location or routing path should be employed.

(7) Mutual authentication

For the secure M2M communication, both machine and server should be trusted. A server should prove that the machine that tries to connect to itself is legal, and the machine should also verify that the server can be trusted. Hence, the process of mutual authentication is required at the initial connection stage, and for this purpose, the third party trust environment that verifies both server and machine may be employed.

6.2 Femto: Home (Evolved) NodeB

6.2.1 General

The term "femto cell" has become a common term in today's society. Literally femto provides smaller cell size, thus bringing focused coverage. Any wireless technology can provide femto coverage but in this chapter we will focus on 3GPP standardization known as Home (evolved) NodeB (H(e)NB). Here HNB is meant for UMTS and HeNB for LTE. A high-level architecture of H(e)NB system is given in Figure 6.1.

As given in Figure 6.1, H(e)NB is expected to be placed in residence or in enterprise to provide mobile operator coverage, thus using the spectrum assigned to the operator. The purpose is either to provide coverage or to provide higher bandwidth at specific locations. H(e)NB will be connected to the mobile network via any Internet connectivity, e.g., ADSL or FTTH, at the location of deployment thus the figure shows insecure network through which the traffic between H(e)NB and the mobile network traverse. At the mobile network end, there is a security gateway (SeGW) that takes care of secure connectivity, secure communication, and authentication of the H(e)NB. Provisioning of connection to network through connectivity available at the operator network

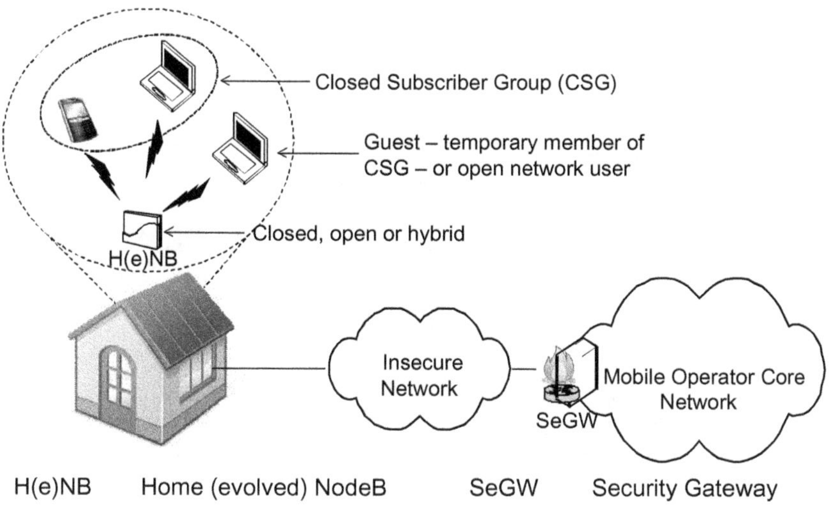

Fig. 6.1 Overview of 3GPP Home (evolved) NodeB technology.

could also mean that the H(e)NB is connecting through some form of modem or router at the customer premises.

The H(e)NB system or technology opens doors for new business possibilities for the mobile operator; one such possibility is shown in Figure 6.1 as the closed subscriber group (CSG). The CSG could be used to provide special services to the members of the given group and administrator of a given group could give access to its H(e)NB to some guests for a given period. One can think of several other services, for the operator H(e)NB also means a way to get the network into the customers location (home, office or anywhere), thus potentially allowing communication business with anything in customer premises. 3GPP system further allows the H(e)NB to be owned by a subscriber who could provide coverage or services at a given location; such subscriber is known as hosting party (HP) and is authenticated by a universal subscriber identity module (USIM) located in H(e)NB.

6.2.2 System Architecture

The system architecture of H(e)NB is depicted in Figure 6.2. The insecure network connecting H(e)NB to the SeGW is known as backhaul link.

The SeGW can be connected to a Home (evolved) NodeB Gateway (H(e)NBGW) or mobility management entity (MME) and is connected to a

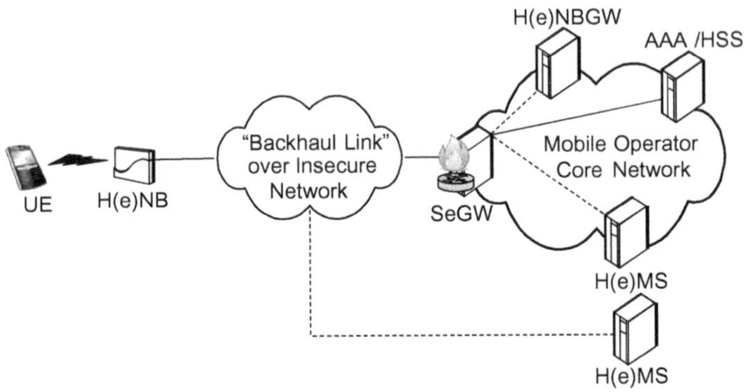

AAA	Authentication, Authorization and Accounting	UE	User Equipment
H(e)MS	Home (evolved) Management System	H(e)NB	Home (evolved) NodeB
HSS	Home Subscriber Subsystem		

Fig. 6.2 H(e)NB system architecture.

AAA/HSS. After mutual authentication, between SeGW and H(e)NB a secure tunnel is created based on IPsec. Any traffic from the H(e)NB to the operator traverses through this tunnel. Of course, an operator can decide not to deploy IPsec but then the operator should provide confidentiality and integrity protection of the traffic.

When it comes to authentication, the H(e)NB has a mandatory device authentication and optional hosting party authentication (HPA) using a hosting party module (HPM) that is basically the USIM. For the purpose of authentication the SeGW must communicate with the AAA server/HSS.

The H(e)NBGW can be either separate or integrated with the SeGW; it performs access control. It is also possible that the HeNBGW is not deployed in which case the MME will perform access control.

The management of H(e)NB is performed by home (evolved) management system (H(e)MS) that can be located either in the operator's network or elsewhere. There should be secure communication between the H(e)NB and the H(e)MS.

6.2.3 Security Threats and Requirements

In this section, we study a few potential threats on H(e)NB and also discuss some of the requirements on the H(e)NB system discussed in the previous section [see 3GPP TS 33.820 and TS 33.320].

From the viewpoint of standardization, it is the first time that a 3GPP network element will be placed in customer premises. This brings the mobile operator's network within reach of the end user. Putting the H(e)NB in customer premises could therefore lead to potential attacks like rogue H(e)NB, man-in-the-middle attack, access or modification of H(e)NB security credentials, modification of the software or firmware in the H(e)NB, and several other potential attacks that can come from both the wired and wireless interfaces. These lead to security requirements as depicted in Figures 6.3 and 6.4 and briefly discussed below:

- When the H(e)NB is switched on or restarted there should be a validation that everything in the device is fine. This is the device integrity validation.
- Although device integrity validation can be done, it is also equally important that the required credentials and parameters are stored such that they are inaccessible to anyone. Therefore there is a necessity of having trusted environment in the H(e)NB.
- The H(e)NB will be the point where (at least some) traffic from the user equipment (UE) will be decrypted, thus making it open for access to potential attackers or just making it available to any other who should not be authorized to access it. Therefore there should be proper means available such that sensitive data is not accessible to anyone in plain text.

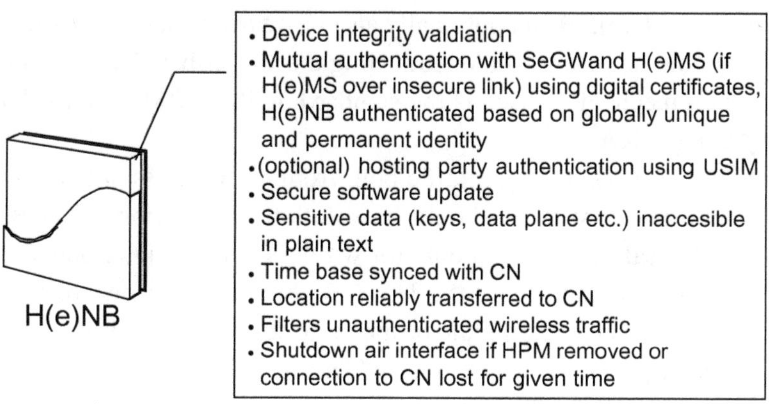

Fig. 6.3 Requirements on H(e)NB.

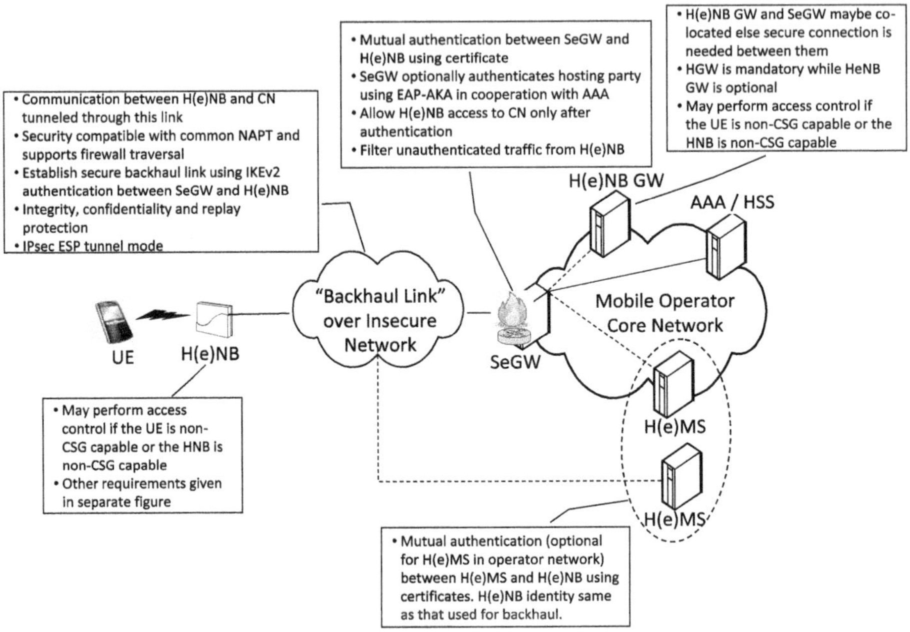

Fig. 6.4 Security requirements on H(e)NB.

- There should also be mutual authentication between the network and the H(e)NB so that both sides are sure about who they are connecting to. The end-point in the network is the SeGW with authentication credentials residing at AAA/HSS. Therefore secure (confidentiality, integrity, and replay protection) communication between the H(e)NB and the SeGW must be possible.
- As the H(e)NB can optionally also be owned by a subscriber, there should be means of authenticating the subscriber. The authentication mechanism used is the standard 3GPP USIM-based solution using AKA.
- The H(e)NB will get traffic both from the wireless and the wired end therefore there should be solution for filtering unnecessary traffic (although given only for wireless in the requirements).
- Every now and then, the H(e)NB software or firmware will be updated thus there should be means to securely provide the updates and also to have secure update. Thus the H(e)NB should be capable of verifying the software or firmware it has received. Therefore

there should be secure communication with the H(e)NB management system (H(e)MS) and also mutual authentication between the H(e)NB and H(e)MS.

- When it comes to CSG there is a necessity of access control that should be done in HNB or HNBGW else in MME.
- The H(e)NB time base should be synced with the network.
- Provisioning location information of the H(e)NB is a requirement for the system therefore the information provisioning must be secured.

Although various requirements are set by 3GPP, in practice there are also security requirements regarding secure implementation, proper testing, hardening of the operating system, access control to management interface, administrative rights etc.

6.2.4 Solutions

Here we present a brief overview of 3GPP H(e)NB solutions, details are available in 3GPP TS 33.320. Note that TS 33.320 does not give details of several solutions instead gives requirements that should be fulfilled or reuses existing solutions from elsewhere.

- *Secure storage*: There are two types of secure storage defined in 3GPP H(e)NB, one being the HPM and the other being the trusted environment (TrE). The HPM is based on universal integrated circuit card (UICC) while for TrE the specification givens requirements on storage of sensitive data and execution of sensitive functions.
- Mutual authentication: Two mutual authentications are described in specification. One of the mutual authentication is of the hosting party using UICC (with USIM) with extensible authentication protocol–authentication and key agreement (EAP-AKA) and the other is device authentication using certificates available at the H(e)NB and SeGW with Internet key exchange version 2 (IKEv2) [RFC 4306].
- Secure tunnel: IKEv2 is used to setup a Internet Protocol security (IPsec) encapsulating security payload (ESP) tunnel between the H(e)NB and SeGW. The H(e)NB initiates the creation of the tunnel

while the SeGW allocates IP address to the H(e)NB on successful authentication. All communication to the H(e)MS goes via the IPsec tunnel but then the H(e)MS can also lie in the Internet. A tunnel using TLS is also possible between the H(e)NB and H(e)MS that also leads to authentication between the two.

- Device integrity and validation: On boot-up the TrE performs integrity check of the device after which the H(e)NB can connect to the network. Implicit device validation is done with successful authentication because failure in integrity check will lead to sensitive data not being accessible for authentication.
- Device authorization: Besides certificate revocation check using Online Certificate Status Protocol (OCSP) [RFC 2560] a AAA-based authorization is also optionally possible. This verification is to check whether the device is allowed to connect to the network and is based on device-authenticated identity.
- Location verification: There are several options of location verification of the H(e)NB. This location verification can be done using the IP address, line identifier, macro cell information, and GPS information. The location information is securely sent to the network.
- Access control: It is important to provide UE access control especially when talking about CSG. The location of access control varies depending whether it is a HNB or HeNB. In the case of HNB and the CSG readiness of the HNB or UE the access control happens at HNB, H(e)NBGW, or SGSN. While for HeNB the access control happens at the MME.

6.3 Multimedia Broadcast and Multicast Service and Group Key Management

6.3.1 Multimedia Group Services

Diverse multimedia services such as IPTV, full duplex video telephony, multimedia advertising, multiparty gaming, video streaming, and web/audio/video conferencing are becoming more and more popular in current wireless communications. However, due to the large bandwidth requirement of most multimedia services, they can be a severe burden on the wireless access links as well as backhaul networks. For this reason, many researchers, vendors, and network operators have tried to develop new architectures that can appropriately deal with the bandwidth-intensive multimedia services.

One of the characteristics of many multimedia services is that they can be delivered through broadcast and multicast service. In the broadcast and multicast service, a group of clients receive the same data from a dedicated server. The researchers have developed transmission technology for multicast and broadcast, and have proposed mechanisms to manage the clients receiving the same data as a group. Standardization bodies have also developed network and service architectures to efficiently support multicast and broadcast. Among them, multimedia broadcast and multicast service (MBMS) has been spotlighted as a representative architecture for multimedia group services and is explained in this chapter.

6.3.2 Multimedia Broadcast and Multicast Service

6.3.2.1 Introduction

In wireless broadcast and multicast, an access point (AP) or a base station transmits the same data to all receivers in a single cell. Due to the shared nature of wireless medium, broadcast and multicast is very natural in wireless communication, requiring the least amount of network resources over the radio access network. MBMS has been developed by 3GPP, maximally utilizing the wireless advantages. Since MBMS can be run over the existing network infrastructures such as GSM, UMTS, CDMA, W-CDMA, LTE, and so on, the cost for deploying a broadcasting system can be greatly reduced. It is expected that the main applications of MBMS are mobile TV and digital multimedia broadcast (DMB) service via an IP network. The MBMS infrastructure provides a communication channel between service providers and users. The channel for MBMS can be operated basically as bidirectional, compared with a unidirectional downlink channel in the present broadcasting system.

MBMS is classified into two modes: a multicast mode and a broadcast mode. The broadcast mode is one that a single sender transmits data to all receivers in a service area, while the multicast mode is one that one sender transmits data to one or more receivers. In broadcast, a service provider can transmit data to all users without users' requests, and the users shall be able to enable or disable the reception of the data in user equipments. On the other hand, the multicast mode is a point-to-multipoint transmission from a single sender to a specific multicast group within a service area. It allows a sender to selectively transmit to cells within the multicast service area so that only the multicast group can receive the data. In the multicast mode, users generally have to subscribe to a corresponding multicast group. Overall, the broadcast mode is different from the multicast mode in the following three aspects: (1) There are no specific requirements to activate the broadcast mode; (2) Users do not need to subscribe to the broadcast service; (3) There is no charging at the MBMS transport service layer. Charging may be made at the MBMS user service layer.

The public land mobile network (PLMN) operator, a user or a third party on their behalf (e.g., company) can create a group for multicast service. Different from the broadcast mode, the MBMS transport service layer may include charging data for the service, but charging can be supported at the user service layer. As well as the broadcast mode, however, the multicast mode does not guarantee the reception of multicast services over the access network. So, higher layer services or applications in MBMS have to provide a mechanism to guarantee the correct data reception.

6.3.2.2 MBMS architecture

Overview We firstly overview MBMS and explain the characteristics of the MBMS architecture. In the MBMS bearer service, the service is provided in the packet switched (PS) domain to deliver IP multicast datagrams to multiple receivers using minimum network and radio resources. The broadcast mode is supported for both EPS and GPRS, but the multicast mode is supported for GPRS. MBMS for EPS supports E-UTRAN and UTRAN, while MBMS for GPRS supports UTRAN and GERAN.

MBMS is implemented by adding a number of new capabilities to existing functional entities of the 3GPP architecture, making new functional entities. Thus, the existing PS domain functional entities (i.e., GGSN, SGSN, MME, E-UTRAN, UTRAN, GERAN, and UE) must be enhanced to offer the MBMS bearer service. In case of an EPS functional entity, MBMS gateway (GW) exists at the edge between the core network (CN) and the broadcast and multicast service center (BM-SC). In the bearer plane, this MBMS service is supposed to provide the delivery of IP multicast datagrams to User Equipments

(UEs) with a specified quality of service. In the control plane, this service offers management mechanisms for the MBMS bearer service activation status of UEs in the case of the multicast mode. In addition, the MBMS service provides the methods for outsourcing the authorization of the MBMS user service to the BM-SC in the case of the multicast mode. It also provides the systems for controlling the session initiation/modification/termination by the MBMS user service and managing the bearer resources for the distribution of MBMS data.

Functional entities of MBMS architecture To provide MBMS bearer services, existing functional entities in 3GPP network (i.e., GGSN, MME, SGSN, eNodeB/RNC/BSC) have to perform several MBMS-related functions and procedures, some of which are dedicated to MBMS. An MBMS-specific functional entity such as BM-SC is supposed to support diverse MBMS user services like service provisioning or data delivery. Figure 6.5(a) shows the GPRS architecture for MBMS, and Figure 6.5(b) shows the EPS architecture for MBMS.

Broadcast-Multicast Service Center (BM-SC) BM-SC performs various functions of MBMS user service provisioning and delivery for content providers. As an entry point for MBMS transmissions by a content provider, BM-SC is supposed to authorize and initiate MBMS bearer services within PLMN. BM-SC can also be used to schedule MBMS transmissions. BM-SC is a basic functional entity which must exist for each MBMS user service.

Figure 6.6 shows the illustration of the BM-SC functional structure consisting of the following subfunctions:

— *Membership function*:

The BM-SC membership function provides authorization for UEs requesting to activate an MBMS service. The membership function may have subscription data of MBMS service users and generate charging records for MBMS service users.

— *Session and Transmission function*:

The BM-SC session and transmission function should be able to schedule MBMS session retransmissions, and label each MBMS session with an MBMS session identifier to allow the UE to distinguish the MBMS session retransmissions. When IP multicast is used for distribution of payload from GGSN to RNC

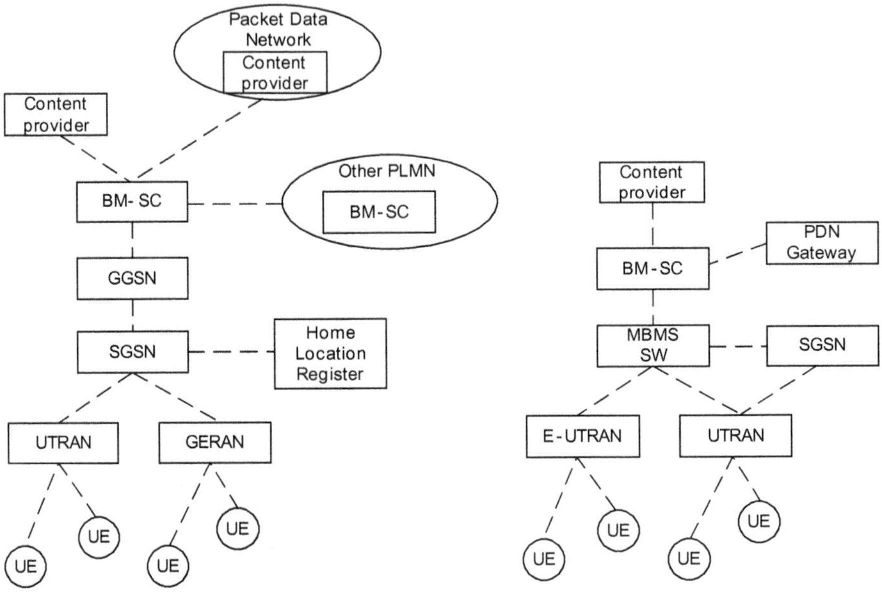

(a) GPRS architecture (b) EPS architecture

Fig. 6.5 MBMS architectures.

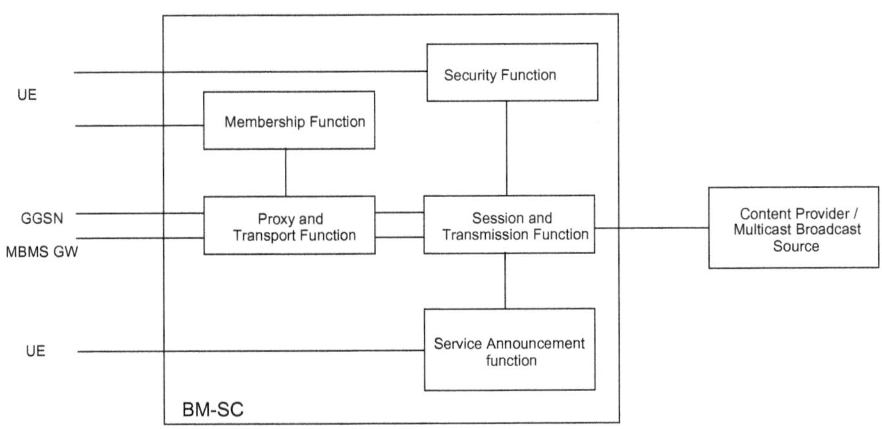

Fig. 6.6 BM-SC functional structure [2].

or from MBMS GW to eNodeB and RNC, the BM-SC session and transmission function is supposed to include synchronization information for the MBMS payload.

— *Proxy and transport function*:

The BM-SC proxy and transport function is a proxy agent for signaling between MBMS GWs/GGSNs and other BM-SC subfunctions. The BM-SC proxy and transport function is supposed to act as an intermediate component for the MBMS data sent from the BM-SC session and transmission function to the MBMS GW or GGSN. In addition, the BM-SC proxy and transport function has the ability to generate charging records for a content provider.

— *Service announcement function*:

The BM-SC service announcement function shall be able to provide service announcements for MBMS user services; media descriptions specifying the media to be delivered as part of an MBMS user service (e.g., type of video and audio encodings); MBMS session descriptions specifying the MBMS sessions to be delivered as part of an MBMS user service (e.g., multicast service identification, addressing, time of transmission, etc.). The media and session descriptions can be delivered by means of service announcements using IETF specified protocols over MBMS bearer services.

— *Security function*:

MBMS user services can utilize the security functions for integrity and/or confidentiality protection of MBMS data. The MBMS security function is used for distributing MBMS keys to authorized UEs.

User equipment In UMTS, user equipment (UE) connects to the base station for MBMS bearer services. The UE has to support the basic functions for activation/deactivation of the MBMS bearer service and the appropriate security functions for MBMS. Once a particular MBMS bearer service is activated, no further explicit user request is required to receive MBMS data.

Depending on UE's capabilities, the UE must be able to receive MBMS user service announcements, non-MBMS-specific paging information and support other services simultaneously. For example, the user can originate or receive a call, or send and receive messages, while receiving MBMS video content. Because the announcements during the reception of paging can create losses in the MBMS data reception, the MBMS user service must be able to appropriately handle such losses. Thus, BM-SC should label each MBMS session with an MBMS session identifier to allow the UE to distinguish the MBMS session retransmissions. In some cases of retransmission, because the UE may already have received the data of the previous MBMS session, the UE can ignore the forthcoming transmission of MBMS session by using the MBMS session identifier.

UTRAN/GERAN Since UTRAN/GERAN has to guarantee the QoS of MBMS data transmission in the multicast mode, the UTRAN/GERAN may require some specific QoS-guarantee mechanisms. For example, prior to or during MBMS transmission, the UTRAN/GERAN chooses an appropriate radio bearer by using the suitable number of users within a cell. In addition, the UTRAN/GERAN has to support the initiation and termination of MBMS transmissions. Since UE's mobility may cause data loss in MBMS data reception, MBMS user services should also be able to handle with potential data loss caused by UE mobility. The UTRAN/GERAN should also have an ability to transmit MBMS user service announcements, non-MBMS-specific paging information and support other services in parallel with MBMS. For example, depending on terminal capabilities, the user can originate or receive a call, or send and receive messages while receiving MBMS services.

SGSN SGSN performs the MBMS bearer service control functions for each individual UE and provides MBMS transmissions to UTRAN/GERAN. SGSN should provide support for intra-SGSN and inter-SGSN mobility procedures. In addition, SGSN can synchronize with the UE. SGSN should also be able to generate charging data per multicast MBMS bearer service for each user.

6.3.2.3 MBMS service provision

In order for users to receive the MBMS bearer services, MBMS needs to be provisioned first, typically by following the specific procedures illustrated in Figure 6.7. The multicast mode is similar to the broadcast mode except for subscription, joining, leaving process, because it serves only the members who subscribe to the service.

— *Subscription*: In a *subscription* phase of the procedure for the service, a user performs a registration process receiving a basic

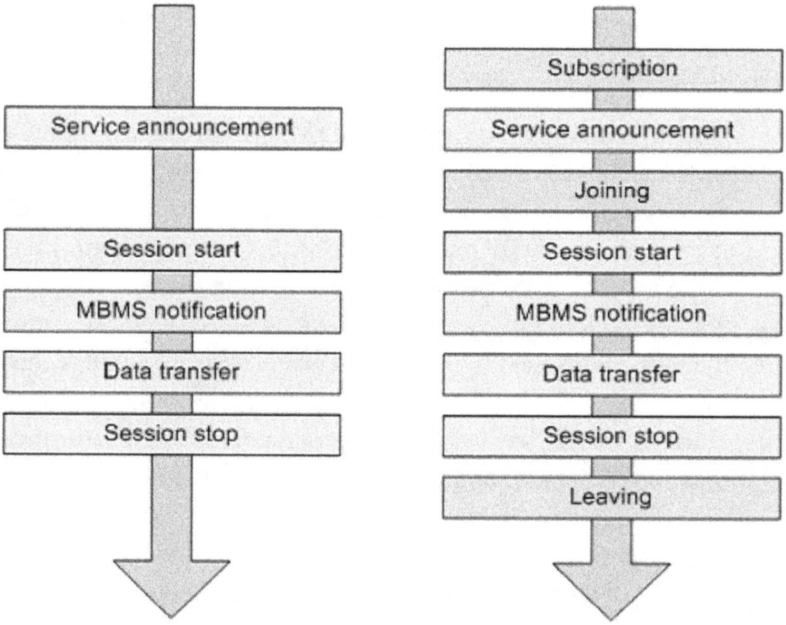

(a) Broadcast mode (b) Multicast mode

Fig. 6.7 Procedures of MBMS service provisioning [2].

registration key to access a service. Registration process may be linked with the charging process. During the procedure, a user is registered to the server (e.g. BM-SC) as a subscribed member. In the service announcement phase, on-going or future service information can be delivered to users. MBMS users can request or receive the information of the range of MBMS user services available. The service information includes service ID (IP multicast address, access point name (APN) address), the information related to service QoS and related parameters (e.g., service start time).

— *Service announcement*: Even though the MBMS specification does not standardize specific MBMS service announcement method, service providers can utilize several service discovery methods for *service announcement* depending on the capability of terminals and the applications that encourage user interrogation.

These methods include web service (HTTP), file transfer protocol (FTP), PUSH mechanism (wireless application protocol (WAP), short messaging service (SMS), multimedia messaging service (MMS)), MBMS multicast mode, MBMS broadcast mode, and SMS cell broadcast. By using the service announcement methods, the service provider can send users the service information including the service ID (IP multicast address), QoS-related information and a schedule of time when service session starts.

— *Joining: Joining* is the procedure that a subscribed user informs the network of the intention to receive a service. Upon receiving the request, the network has to prepare service transmission. The contexts for the service may be generated in SGSN covering the location of the user and will be delivered only to the SGSN.

  The joining time is chosen by the user or UE possibly in response to Service Announcement. Users are supposed to typically join at the time of their choosing, and thus the period between announcement and joining can be very long or very short. In order to avoid the overload situations caused by quite a number of users attempting to join in a short period of time, the UE can use parameters sent by BM-SC in the service announcement to randomize the joining time.

  Some MBMS bearer services can always be on. In this case, Joining can occur immediately after Service Announcement and possibly many hours before or after Session Start. In other cases, if a Session Start time is known, Joining can occur immediately before Session Start or after Session Start. For these services, the announcement can include some indication of a time period that users or UEs can utilize to choose a time for joining the MBMS bearer service.

— *Session Start*: *Session start* informs that a dedicated service is prepared for data transmission. By transmitting the MBMS session start message, BM-SC notices that data may be transmitted soon and triggers the bearer resource establishment for MBMS data transfer. Session Start transmits the data to the only SGSN where joining users exist.

Even if Session Start indicates that the transmission is about to start, the time delay between a Session Start indication and actual data transmission can be long. This is due to the network actions required at Session Start, e.g., provisioning of service information to the UTRAN (i.e., establishment of the bearer plane).

The BM-SC Session can trigger Session Start by using an explicit notification. In the case that bearer plane resources are set-up after the start of session data transmission, the network is not required to buffer the session data but loss of data can be occurred.

— *MBMS Notification*: *MBMS notification* informs each UE of forthcoming (and potentially ongoing) MBMS multicast data transfer and transmits the resource information. To obtain UEs' location information, RNC notifies Session Start to UEs by utilizing group paging. UEs can send a response message to RNC so that RNC can assign necessary network resources. RNC can allocate point-to-point dedicated channels to the cell according to the number of UEs or point-to-mulitpoints channels.

— *Data Transfer*: Data transfer is the procedure that MBMS service data are transferred to the UEs. If data transfer is suspended for more than a certain period of time, BM-SC can send a Session Stop messages for efficient resource management. When receiving the Session Stop message, RNC can release the network resources for the services and allocate the network resources to other services.

— *Session Stop*: It is the point when BM-SC determines that there will be no more data to send for some period of time. With Session Stop, the bearer resources associated with the session are supposed to be released.

— *Leaving*: Leaving is the process that MBMS multicast deactivation is done by the user, by which a subscriber leaves a multicast group. Thus, Leaving message is transmitted when a user no longer wants to receive the service of a specific MBMS bearer service. When a user wants to stop receiving the service, the user can do service deactivation and leave the service that the user has joined.

6.3.2.4 MBMS security

In comparison to the broadcast mode, the multicast mode has to provide a charging function so that only authorized users are able to access multicast services. Because charging is related to security, for the access control of MBMS, data has to be encrypted, and the ciphering key has to be delivered only to the authorized users who have paid or are willing to pay to a service provider of MBMS.

The application layer takes charge of ciphering process for MBMS, and there is no ciphering function in a RAN layer. The MB-SC application layer generates a ciphering key and delivers the key to the authorized users. Since the key can be continually updated, the MB-SC application layer has to provide a key management scheme to generate and deliver a new key to the users, and adopt the key to MBMS services. The new key or information to generate a new key is supposed to be securely transmitted from BM-SC to devices in an individual method.

To adopt a new key, the MB-SC application layer transmits the ciphering key information as a type of additional data with the MBMS service contents. The ciphering key information includes the ID of the ciphering key and additional information to generate a new key. By using the ID information, devices can generate a new key to access the MBMS service contents.

6.3.3 Group Key Management

6.3.3.1 Introduction

In the previous section, we have discussed MBMS services and their service architecture. We note that contrary to the broadcast mode, in the multicast mode, the transmitted data from a server are delivered only to the subscribed users over the network. If a user wants to receive a specific multimedia service, the user has to subscribe to the service. In other words, the multicast mode requires an access control mechanism in order that the only subscribed users can successfully access the data.

As a way of access control, service providers have usually employed some cryptographic techniques, thereby guaranteeing data confidentiality. Data confidentiality implies nongroup confidentiality, which means that no other users except subscribed users should be able to access any data. Nongroup confidentiality includes two cases of confidentiality, i.e., forward confidentiality and backward confidentiality. Forward confidentiality means that the users who left the group should not be able to access any future data. On the other hand, backward confidentiality means that users who have newly joined the group should not be able to access old data.

In enforcing access control for a group of users, a group key plays an essential role in encrypting and decrypting data. A group key must be shared only within subscribed users and must not be revealed to users outside the group including the user who has just left the group. Since group users can leave and join the group whenever they want, the group key should always be managed and updated, whenever the membership of users changes. Thus, it is quite important to understand the mechanism of group key management, which will be explained in the following sections.

One thing we would like to point out is that as we discussed in the previous section, the MBMS application layer takes charge of managing the group key for data security. The standardization organizations do not specify the group key management in detail. That means that the service providers who want to deliver the multicast broadcast services over the various communication networks including IEEE 802.16e, GPRS, digital subscriber line (DSL), and 3GPP MBMS have to adopt separately a group key management mechanism for access control or subscription charging.

In the case of multicast broadcast services (MBS) of IEEE 802.16e, the MBS architecture provides a key state machine as a key management mechanism, but it does not provide the detailed algorithms for efficiently managing a group key. The specification of MBS defines only the key management in layer 2, saying that "The generation and transport of the MAK (MBS AK) is outside the scope of the IEEE 802.16 standard. It is provided through means defined at higher layers." Although the detailed group key management mechanism and related algorithms are mainly out of the scope of standards, we briefly overview some important research results for the interested readers in the following subsection.

6.3.3.2 Group Key Management

In typical group key management, a trusted third party known as key distribution center (KDC) is employed for access control, having an authority

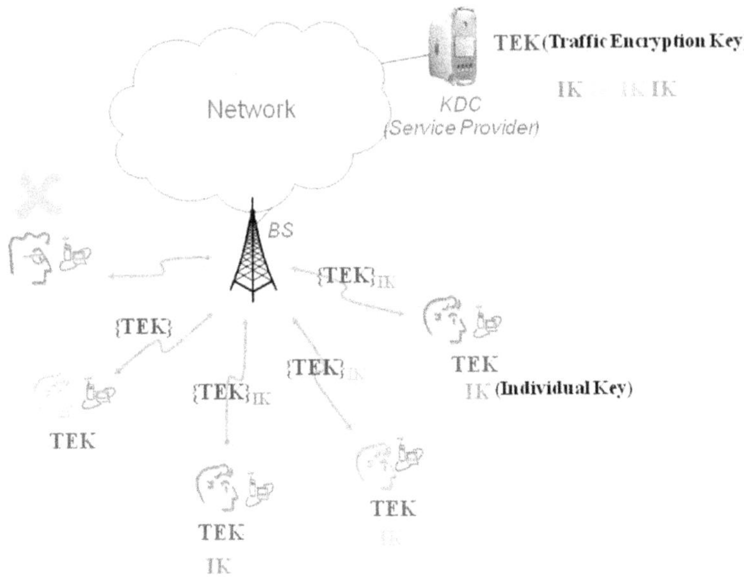

Fig. 6.8 Illustration for group key management.

of managing a group key. Usually, a service provider or a network operator manages the KDC. Basically, the KDC is assumed to have exchanged the individual key (IK) in advance with each subscribed user. Whenever the KDC needs to update the group key, the KDC encrypts a new group key with each subscribed user's IK to provide the data confidentiality during key distribution, and transmits the encrypted new group key to all subscribed users, respectively. Figure 6.8 describes the concept of group key management.

6.3.3.3 Key tree structure

Because the KDC always has to transmit the messages containing a new group key to subscribed users individually whenever the KDC needs to update a group key, the number of key update messages is proportional to the number of MBMS service users. As the number of MBMS subscribers increases, key update messages can be heavy load to a network. Generally, this overhead is called the scalability problem of group key management.

To alleviate the scalability problem, many group key management protocols have been proposed so far. The protocols are mainly based on the idea that the KDC divides group members into subgroups. The KDC assigns subgroup

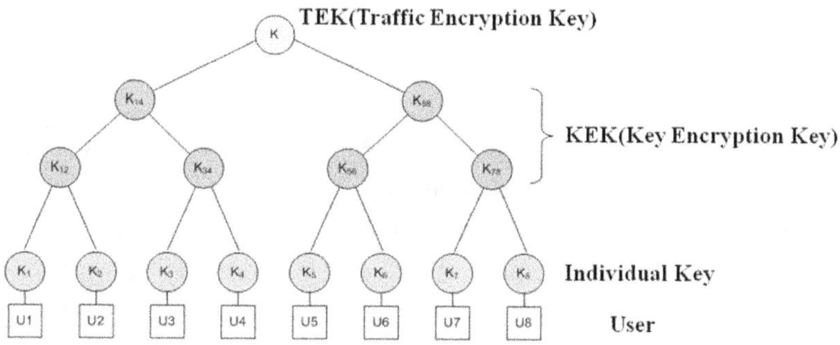

Fig. 6.9 Illustration of a key tree struture.

keys to each subgroup so that they are shared only by the subgroup users. Subgroups can further be divided into small subgroups that have their own subgroup keys. Thus, the group key can be managed with a key tree structure constructed by layered subgroup keys. With a key tree structure, the KDC can efficiently deliver secure data (a new group key or subgroup keys) to specific subgroup users. As a result, a key tree structure reduces the number of update messages by the logarithmic order of the number of MBMS users (from O(N) to O(logN)).

Figure 6.9 illustrates an example of a key tree structure. The traffic encryption key (TEK) is a group key to encrypt and decrypt data, and key encryption key (KEK) is a subgroup key to encrypt and decrypt a group key or upper level subgroup keys. Let us assume that a group key K needs to be updated and replaced by a new group key K'. Without the key tree structure, the KDC has to transmit the eight messages as follows:

$$\text{KDC} \rightarrow \text{Users: } M_i = \text{Enc}(K', K_i) \text{ for } i = 1,2,\ldots,8.$$

Here, M_i is an encrypted message and Enc(K', K) means that data K' is encrypted with key K. With the key tree structure, on the other hand, the KDC needs to transmit only two messages as follows:

KDC $\rightarrow$ U1 $\sim$ U4: $M_{14} = \text{Enc}(K', K_{14})$;

KDC $\rightarrow$ U5 $\sim$ U8: $M_{58} = \text{Enc}(K', K_{58})$.

6.3.3.4 Batch rekeying

Besides the key tree structure, batch rekeying also helps alleviate the scalability problem. For strict data confidentiality, the KDC must update a group key whenever user membership is changed, which is called individual rekeying. Batch rekeying, on the other hand, updates a group key once in a time interval after the KDC collects join or leave requests during the time duration.

Batch rekeying greatly reduces the scalability problem because it can decrease the number of rekeying events, which in turn reduces the number of messages. For example, let's assume that U1 and U3 leave a MBMS group supported by the key tree structure in Figure 6.9. Individual rekeying requires ten key update messages as follows. Here, we assume that the key tree structure is maintained.

When U1 leaves

K, K_{14}, K_{12} must be updated to K', K'_{14}, K'_{12}, respectively

KDC $\rightarrow$ U2 $\sim$ U4: $M_{14} = Enc(K', K'_{14})$,

KDC $\rightarrow$ U5 $\sim$ U8: $M_{58} = Enc(K', K_{58})$,

KDC $\rightarrow$ U2: $M_{12} = Enc(K'_{14}, K'_{12})$,

KDC $\rightarrow$ U3 $\sim$ U4: $M_{34} = Enc(K'_{14}, K_{34})$,

KDC $\rightarrow$ U2: $M_2 = Enc(K'_{12}, K_2)$.

When U3 leaves

K', K'_{14}, K_{34} must be updated to K", K''_{14}, K'_{34}, respectively

KDC $\rightarrow$ U2, U4: $M'_{14} = Enc(K'', K''_{14})$,

KDC $\rightarrow$ U5 $\sim$ U8: $M'_{58} = Enc(K'', K_{58})$,

KDC $\rightarrow$ U2: $M'_{12} = Enc(K''_{14}, K'_{12})$,

KDC $\rightarrow$ U4: $M'_{34} = Enc(K''_{14}, K'_{34})$,

KDC $\rightarrow$ U4: $M_4 = Enc(K'_{34}, K_4)$.

Batch rekeying, however, requires only six key update messages as follows since the group keys are updated simultaneously:

K, K_{14}, K_{12}, K_{34} must be updated to K', K'_{14}, K'_{12}, K'_{34}, respectively

KDC $\rightarrow$ U2, U4: $M_{14} = Enc(K', K'_{14})$,

KDC $\rightarrow$ U5 $\sim$ U8: $M_{58} = Enc(K', K_{58})$,

KDC $\rightarrow$ U2: $M_{12} = Enc(K'_{14}, K'_{12})$,

KDC $\rightarrow$ U4: $M_{34} = Enc(K'_{14}, K'_{34})$,

KDC $\rightarrow$ U2: $M_2 = Enc(K'_{12}, K_2)$.

KDC $\rightarrow$ U4: $M_4 = Enc(K'_{34}, K_4)$.

Even though batch rekeying greatly reduces the scalability problem, batch rekeying may sacrifice forward confidentiality, an essential requirement of access control. In batch rekeying, even though a user leaves a multicast group, the user can still access future data until the KDC updates a group key. On the contrary, individual rekeying guarantees the perfectly secure multicast service because it is supposed to update a group key right after the KDC is notified a group membership change.

6.4 Conclusion

In this section, we reviewed MBMS and group key management to support group-based services. We discussed the MBMS architectures, securities, and charging mechanisms. We also overviewed the key management scheme for group-based multicast services. Since MBMS is expected to be popular for multimedia services, group key management will also be an efficient access control mechanism for MBMS.

References

M2M:

[1] 3GPP TS 22.368, Service requirements for Machine-Type Communications (MTC); (Stage 1).
[2] 3GPP TR 23.888, System improvements for Machine-Type Communications (MTC).
[3] 3GPP TR 33.868, Security aspects of Machine-Type Communications.

H(e)NB:

[4] 3GPP TS 33.320, 3G Security; Security of Home Node B (HNB)/Home evolved Node B (HeNB).
[5] IETF RFC 4187, Extensible Authentication Protocol Method for 3rd Generation Authentication and Key Agreement (EAP-AKA).
[6] IETF RFC 4306, Internet Key Exchange (IKEv2) Protocol.
[7] IETF RFC 4739, Multiple Authentication Exchanges in the Internet Key Exchange (IKEv2 Protocol, November 2006.
[8] The Broadband Forum TR-069, CPE WAN Management Protocol v1.1, Issue 1 Amendment 2, December 2007.
[9] IETF RFC 4346, The Transport Layer Security (TLS) Protocol Version 1.1.
[10] IETF RFC 5246, The Transport Layer Security (TLS) Protocol Version 1.2.
[11] IETF RFC 2560, X.509 Internet Public Key Infrastructure Online Certificate Status Protocol — OCSP.
[12] IETF RFC 4806, Online Certificate Status Protocol (OCSP) Extensions to IKEv2.

MBMS:

[13] 3GPP TS 22246-900, 3rd Generation Partnership Project; Technical Specification Group Services and System Aspects; Multimedia Broadcast/Multicast Service (Stage 1).
[14] 3GPP TS 23246-950, 3rd Generation Partnership Project; Technical Specification Group Services and System Aspects; Multimedia Broadcast/Multicast Service (MBMS); Architecture and functional description.
[15] 3GPP TS 33.102, 3rd Generation Partnership Project; Technical Specification Group Services and System Aspects; 3G Security; Security Architecture.
[16] LAN/MAN Standards Committee, P802.16Rev2/D1 Part 16: Air Interface for Broadband Wireless Access Systems.

Abbreviations

3DES	Triple DES
3GPP	Third Generation Partnership Project
AAA	Authentication, authorization, and accounting
AAAH	AAA Home
ACL	Access control list
ADSL	Asynchronous digital subscriber line
AES	Advanced encryption standard
AK	Authorization key
AKA	Authentication and key agreement
AKID	AK identity
AN	Access network
AP	Access point
APN	Access point name
AS	Application server
AS	Access stratum
ASME	Access security management entity
ASN	Access service network
ASN-GW	Access service network gateway
ATM	Asynchronous transfer protocol
AuC	Authentication center
AUTN	Authentication token
AVP	Attribute value pairs
AWG	Application working group
BM-SC	Broadcast multicast service center
BRH	Bandwidth request header

BS	Base station
BSC	Base station controller
BTS	Base transceiver station
CBC	Cipher block chaining
CBC-MAC	Cipher block chaining message authentication code
CCM	Counter with CBC-MAC
CDMA	Code division multiple access
CHAP	Challenge authentication protocol
CI	CRC indicator
CID	Connection identifier
CINR	Carrier-to-interference-and-noise ratio
CLF	Connectivity session and repository location function
C-List	Compatibility list
CMAC	Cipher-based message authentication code
CN	Core network
CODEC	Coder decoder
CPS	Common part sublayer
CPU	Central processing unit
CRC	Cyclic redundancy check
C-RNTI	Cell-RNTI
CS	Circuit switched
CS	Convergence sublayer
CSCF	Call session control function
CSG	Closed subscriber group
CSN	Connectivity service networks
CT	Core networks and terminals
C-TEID	Common TEID
C-TEID	Common TEID
CTR	Counter
CWG	Certification Working Group
DDoS	Distributed DoS
DES	Data encryption standard
DES-CBC	Date encryption standard-cipher block chaining
DH	Diffie Hellman
DL-MAP	Down link MAP
DMB	Digital multimedia broadcast

DNS	Domain name system
DoS	Denial of service
DRM	Digital rights management
DSL	Digital subscriber line
EAP	Extensible authentication protocol
EAP-TTLS	EAP tunneled transport layer security
EARFCN-DL	E-UTRA absolute radio frequency channel number-down link
EC	Encryption control
EDGE	Enhanced data rate for GSM evolution
EEA	EPS encryption algorithm
EIA	EPS integrity algorithm
EIK	EAP integrity key
EIR	Equipment identity register
EKS	Encryption key sequence
EMM	EPS mobility management
EMSK	Extended master session key
EMSK	Extended master session key
eNB	evolved NodeB
ENIAC	Electrical numerical integrator and calculator
EPC	Evolved packet core
ePDG	evolved packet data gateway
EPS	Evolved packet system
ES	Emergency services
ESF	Extended subheader field
ESM	EPS session management
ESP	Encapsulating security payload
ETWG	Evolutionary Technical Working Group
E-UTRAN	Evolved-UTRAN
FBSS	Fast base Station switching
FDD	Frequency division duplex
FTP	File transfer protocol
FTTH	Fiber to the home
GERAN	GSM EDGE radio access network
GERAN	GSM EDGE radio access network
GGSN	Gateway GPRS support node

GKEK	Group key encryption key
GMH	Generic MAC (medium access control) header
GPRS	General packet radio service
GRWG	Global Roaming Working Group
GSA	Group Security Association
GSM	Global system/standard for mobile communications
GTEK	Group traffic encryption key
GTP-U	GPRS tunneling protocol for user plane
GUMMEI	Globally unique MME identifier
GUTI	Globally unique temporary identity
GW	Gateway
H(e)MS	Home (evolved) management system
H(e)NB	Home (evolved) NodeB
HA	Home agent
HARQ	Hybrid automatic repeat request
HCS	Header check sequence
HDMO	Macro diversity handover
HeNBGW	Home evolved NodeB gateway
HHO	Hard handover
HLR	Home location register
HMAC	Hash-based message authentication code
HN	Home network
H-NSP	Home-NSP
HO	Handover
HP	Hosting party
HSPA+	High-speed packet access plus
HSS	Home subscriber server
HSS	Home subscriber subsystem
HT	Header type
HTTP	Hyper text transfer protocol
IC	Integrated circuit
I-CSCF	Interrogating CSCF
ID	Identity
IDEA	International data encryption algorithm
IE	Information element

IEEE	Institute of Electrical and Electronic Engineering
IETF	Internet Engineering Task Force
IK	Individual key (for Chapter 6 on MBMS)
IKE	Internet key exchange
IM	IP multimedia
IMEI	International mobile station equipment identity
IMEISV	IMEI and software version number
IMPI	IM private identity
IMPU	IM public identity
IMS	IP multimedia subsystem
IMSI	International mobile subscriber identity
IMT	International mobile telecommunications
IP	Internet Protocol
IPsec	Internet Protocol security
ISIM	IM services identity module
ISR	Idle state signaling reduction
ITU	International Telecommunications Union
IV	Initialization vector
KDC	Key distribution center
KDF	Key derivation function
KEK	Key encryption key
KSI	Key set identifier
L1/L2	Layer 1/2
L2TP	Layer 2 tunneling protocol
LBS	Location-based service
LEAP	Lightweight EAP
LEN	Length
LI	Lawful intercept
LSM	Limited service mode
LTE	Long-term evolution
M2M	Machine-to-machine
MAC	Message authentication code
MAC	Medium access control
MBMS	Mobile broadcast and multicast services
MBRA	Multicast and broadcast rekeying algorithm

MBRA	Multicast and broadcast rekeying algorithm
MBS	Multicast and broadcast
MBSAK	Multicast and broadcast service authorization key
MCBCS	Multicast and broadcast services
MCC	Mobile country code
MCIM	Machine communication identity module
MD5	Message digest 5
MDHO	Macro diversity handover
ME	Mobile equipment
MIME	Multipurpose Internet mail extensions
MIMO	Multiple input multiple output
MIP	Mobile IP
MK	Master key
MME	Mobility management entity
MMEC	MME code
MMEGI	MME group ID
MMS	Multimedia messaging service
MNC	Mobile network code
MPDU	MAC protocol data unit
MS	Mobile station
MSC	Mobile switching center
MS-CHAP	Microsoft CHAP
MSDU	Mac service data unit
MSIN	Mobile subscriber identity number
MSK	Master session key
MTK	MBS traffic key
M-TMSI	MME-TMSI
MTU	Maximum transmission unit
MWG	Marketing Working Group
NAP	Network access provider
NAS	Network access server
NAS	Nonaccess stratum
NB	NodeB
NCC	Next hop chaining counter
NesCom	New Standards Committee

NGMN	Next generation mobile network
NH	Next hop
NMSI	National mobile subscriber identity
NRM	Network reference model
NSP	Network service provider
NVM	Nonvolatile memory
NWG	Network working group
OAM&P	Operation, administration, maintenance, and provisioning
OTA	Over-the-air
OTP	One-time password
PAK	Primary AK
PAP	Password authentication protocol
PAR	Project authorization request
PCC	Policy and charging control
PCG	Project coordination group
PCI	Physical cell ID
PCRF	Policy and charging rules function
PDCP	Packet data control protocol
PDG	Packet data gateway
PDN	Packet data network
PDN	GW or PGW Packet data network gateway
PDP	Packet data protocol
PDU	Packet data unit
PEAP	Protected EAP
PEM	Privacy enhanced mail
PGP	Pretty group privacy
PGW	Packet data network gateway
PHY	Physical layer
PIN	Personal identification number
PK	Public key
PKI	Public key infrastructure
PKM	Privacy key management
PLMN	Public land-mobile network
PMK	Pairwise master key
PMP	Point-to-multipoint

PN	Packet number
PPP	Point-to-point protocol
PPTP	Point-to-point tunneling protocol
Pre-PAK	Pre-primary AK
PRF	Psuedorandom function
PS	Packet switched
PSK	Preshared key
QoE	Quality of experience
QoS	Quality of service
RADIUS	Remote authentication dial-in user service
RAN	Radio access network
RAT	Radio access technology
RAU	Routing area update
RC4/5	Rivest's Cipher 4/5
REQ	Request
RevCom	Review Committee
RF	Radio frequency
RLC	Radio link control
RNC	Radio network controller
RNG	Random number generator
RNTI	Radio network temporary identity
RP	Reference point
RRC	Radio resource control
RSA	Rivest, Shamir and Adleman
RSP	Response
RSS-CN	Received signal strength
RSSI	Received signal strength index
RWG	Regulatory Working Group
S	Start
S/MIME	Secure MIME
S1-AP	S1 interface application protocol
SA	Security association
SA	Service and systems aspect (of 3GPP standards)
SAE	System architecture evolution
SAID	Security Association ID

SAP	Service access point
SAx	SA WGx
SBC	SS basic capability
SCTP	Stream control transmission protocol
SeGW	Security gateway
SGSN	Serving GPRS support node
SGW	Serving gateway
SI	Study item
SID	Study item description
SIM	Subscriber identity module
SLA	Service level agreement
SMC	Security mode command
SMS	Short message service
SMSC	Short message service centre
SNID	Serving network identity
SNR	Serial number
SPWG	Service Provider Working Group
SRVCC	Single radio voice call continuity
SS	Subscriber station
SS7	Signalling system 7
SSL	Secure socket layer
SVN	Software version number
TA	Tracking area
TAC	Tracking area code
TAC	Type allocation code
TAI	Tracking area identity
TAU	Tracking area update
TDD	Time division duplex
TDMA	Time division multiple access
TEK	Traffic encryption key
TLS	Transport layer security
TLS-PRF	TLS Pseudorandom function
TLV	Threshold limit falue
TMSI	Temporary mobile subscriber identity
TR	Technical report

TrE	Trusted environment
TS	Technical specification
TSG	Technical Specification Group
TTLS	Tunneled TLS
TWG	Technical Working Group
UDP	Universal datagram protocol
UE	User equipment
UL-MAP	Up Link MAP
UMTS	Universal mobile terrestrial system
USI	Universal services interface
USIM	Universal subscriber identity module
UTRA	Universal terrestrial radio access
UTRAN	UMTS terrestrial radio access network
VCC	Voice call continuity
VLR	Visited location register
V-NSP	Visited-NSP
VPN	Virtual private network
WAG	Wireless access gateway
WAP	Wireless access protocol
W-CDMA	Wideband code division multiple access
WGx	Work Group × (1, 2,...)
WI	Work item
WID	Working item description
WiMAX	Worldwide interoperability for microwave access
WLAN	Wireless local area network
WMAN	Wireless metropolitan area network
WRAP	Wireless router application platform (?)
xDSL	Extended digital subscriber line

Index

RIVER PUBLISHERS SERIES IN STANDARDISATION

Other books in this series:

Future Trends and Challenges for ICT Standardization
ISBN: 978-87-92329-38-7

Entity Authentication and Personal Privacy in Future Cellular Systems (Errata Available)
ISBN: 978-87-92329-32-5

Wireless Independent Living for a Greying Population
ISBN: 978-87-92329-22-6

For Product Safety Concerns and Information please contact our EU representative GPSR@taylorandfrancis.com
Taylor & Francis Verlag GmbH, Kaufingerstraße 24, 80331 München, Germany

www.ingramcontent.com/pod-product-compliance
Lightning Source LLC
Chambersburg PA
CBHW071243230926
51849CB00001B/215
* 9 7 8 8 7 9 2 3 2 9 6 3 9 *